China Handbook Series

EDUCATION
AND SCIENCE

FOREIGN LANGUAGES PRESS   **BEIJING**

CHINA HANDBOOK SERIES

EDUCATION AND SCIENCE

Compiled by
the *China Handbook* Editorial Committee

Translated by
Zhou Yicheng, Cai Guanping and Liu Huzhang

FOREIGN LANGUAGES PRESS   BEIJING

First Edition   1983

ISBN   0-8351-0988-7

Published by the Foreign Languages Press
24 Baiwanzhuang Road, Beijing, China

Printed by the Foreign Languages Printing House
19 West Chegongzhuang Road, Beijing, China

Distributed by China Publications Centre (Guoji Shudian)
P.O. Box 399, Beijing, China

Printed in the People's Republic of China

EDITOR'S NOTE

More than 30 years have elapsed since the birth of the People's Republic of China on October 1, 1949. "What is China really like today?" many people abroad wish to know. To answer this question, we plan to compile and publish a voluminous *China Handbook*, in which we intend to introduce the New China in every field of its activities. Emphasis will be on the process of development during the past three decades, the accomplishments, and the problems that still remain. The book will contain accurate statistics and related materials, all of which will be ready references for an interested reader.

To enhance the usefulness of the forthcoming volume, we plan to publish 10 major sections separately at first, so that we shall have an opportunity to take into consideration the opinions of our readers before all the composite parts are put together, revised and published as one volume. These separate sections are:

Geography
History
Politics
Economy
Education and Science
Literature and Art
Sports and Public Health
Culture

Life and Lifestyles

Tourism

Here, we wish particularly to point out the following:

First, the statistics listed in each separate book exclude those of Taiwan, unless otherwise indicated.

Second, the statistics are those compiled up to the end of 1980.

The *China Handbook* Editorial Committee

CONTENTS

Beijing University, whose predecessor is the Metropolitan College, founded in 1898.

Classroom building of Qinghua College, predecessor of Qinghua University. Completed in 1911 with Boxer Indemnity funds, the building has been preserved.

Founded in 1921, Xiamen (Amoy) University is well-known in south China

Students at Qinghua University in optical experimentation.

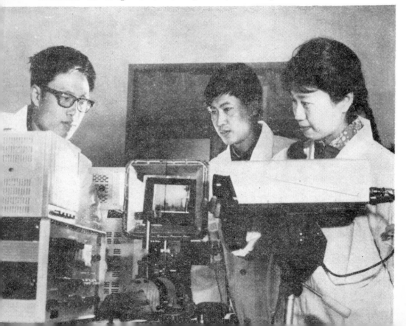

A chemistry laboratory in Fudan University, Shanghai.

Uygur students in the Northwest China Institute for Nationalities.

Students at a class of w[...]
dyeing in the Central Ins[...]
tute of Arts and Cra[...]
Wax-dyeing is an ancie[...]
Chinese art.

Chinese and foreign students doing *taijiquan* (shadow boxing), which has a history of more than 300 years in China. It strengthens the body and has certain therapeutic effects.

Foreign students in a Chinese college.

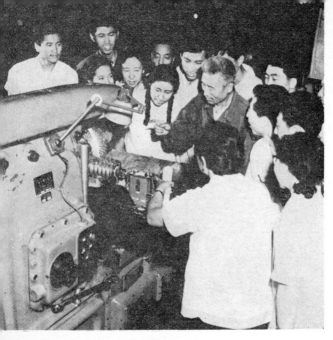

A teacher at Harbin Workers' Spare-Time College gives an on-the-spot lesson.

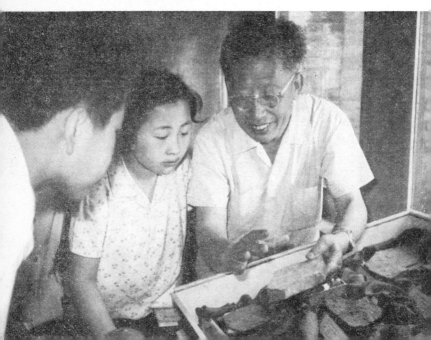

Middle school students learn something about agronomy and other knowledge after regular classes.

A primary school teacher referees her pupils' tug-of-war.

Kindergarten scene.

Chapter One

EDUCATION

1. EDUCATION IN ANCIENT AND MODERN CHINA

(1) EDUCATION IN ANCIENT CHINA

China is one of the oldest countries that has an educational system. Schools came into existence as early as the Western Zhou (c. 11th century to 770 B.C.). Among the subjects taught were rites (ritual and rules), music (ceremonial music and ritual dances), archery, chariot-riding, history and mathematics. Jointly these subjects were referred to as the "six arts". The government ran and controlled the schools; government officials and teachers were in fact the same people.

Private education began to make its appearance during the Spring and Autumn and the Warring States period (770-221 B.C.). Confucius (551-479 B.C.) was a pioneer in this regard. Reportedly he had altogether 3,000 disciples, of whom 72 became proficient in the "six arts". Besides Confucius, there were such famous private educators as Mo Zi, Mencius and Xun Zi. History shows that by the time of the Western Han (206 B.C.-A.D. 24) private and public education existed side by side as the basic forms of feudal education.

Public schools were those established by governments at different levels. There were central schools under the direct control of the central government and local schools administered by local authorities. Emperor Wu Di (r. 141-87 B.C.) of the Han Dynasty established an Imperial College as the highest institution of learning for the propagation of Confucian classics. Teachers there were scholars knowledgeable in the "Five Classics".* With schools known as *xue* (学), *xiao* (校), *yang* (庠) and *xu* (序) being set up in prefectures, counties, districts and townships, a network of education was set up across the country. Succeeding dynasties, to a great extent, did the same.

Ancient education reached a most glorious stage during the Sui and Tang dynasties (A.D. 581-907). Besides the imperial college, there were other central-administered schools to educate the children of noblemen and high officials, as well as schools specializing in calligraphy, mathematics, law (including imperial edicts and decrees), and medical science. On the local level, there were schools to teach Confucian classics and medical science, administered by prefectures, subprefectures and counties. During the Ming Dynasty (A.D. 1368-1644) military schools were established by both the central and local authorities. Beginning in the Tang Dynasty there was in the central government an organ known as Guo Zi Jian (Department of Cultural Affairs) responsible for the administration of public schools. During the Song Dynasty (A.D. 960-1279), Guo Zi Jian also functioned as the na-

---

*The "Five Classics" are the *Book of Odes*, *Book of Documents*, *Book of Rites*, *Book of Change* and *The Spring and Autumn Annals*.

tion's highest institution of learning. Public education had thus become more complete.

During the Han Dynasty, great advances were made in private education. Because of a limited number of public schools, children from landlord families generally enrolled in private schools. Known sometimes as *xue guan* (study halls) or *shu guan* (halls of learning), these private schools were of two kinds, "elementary" and "high". In the elementary schools, children learned to read and write, whereas in high schools, which corresponded to modern colleges, students studied the Confucian classics, such as the "Four Books"* and the "Five Classics". After the 10th century, printing by wooden blocks had advanced to such an extent that printed textbooks, known as "primers", became very popular. Among them the *Three-Character Classics*, *One Hundred Surnames* and *One Thousand Characters* were most widely circulated. After the 14th century, academies emerged, and private schools were primarily concerned with elementary education. Then there were these kinds of private schools. First, there were the "teaching halls" or "sitting halls" where children of landlords and merchants were taught in their own homes by invited teachers. Second, there were the "family halls" where pupils were taught at teachers' homes. Third, there were the "charity halls" or "charity schools" located in such public places as temples and shrines, funded by the local gentry and attended by children of the common people. Still students studying in private schools, like those studying in public schools, were mostly children of a rich family background, and enrollment from families of the poor was extremely rare.

---

* The "Four Books" are *The Analects of Confucius*, *Mencius*, *The Great Learning* and *The Doctrine of the Mean*.

During the Tang Dynasty, academies had been the place where books were collected and edited; only later did they acquire the characteristics of schools. By the Song Dynasty, an increasing number of schools of this kind had appeared. Run mostly by "famous scholars" and emphasizing "free discussion", they had a student body drawn exclusively from families of the feudal gentry and landlords. They used the method of collective discussion to study Confucian classics. Originally, they had been run by private individuals; gradually they came under the control of the government and were run on a semi-official basis. By the Qing Dynasty (1644-1911) when individuals were prohibited from having anything to do with private education, all academies became governmental establishments.

Beginning in the Sui-Tang period, the civil service examination had a great impact on education. Through examination, the feudal rulers selected government officials, and Confucian classics, such as the "Four Books" and the "Five Classics", formed the core of the subjects tested. On the one hand, the examination system strengthened central autocracy and provided a portion of intellectuals from medium and small landlord families with the opportunity to enter into officialdom. On the other hand, it degraded education and made it a mere appendage to the examination system. This was especially true beginning in the Ming Dynasty when the writing of the "eight-legged essay" (a prescribed form set to a fixed pattern of style with a limited number of words to elucidate the meaning of Confucian classics) was taken as a criterion for passing the examination.

(2) EDUCATION IN MODERN CHINA

Following the invasion of capitalist forces from the West during the late Qing period, Britain, the United States, and other countries established missionary schools in the coastal areas of China and began to put into practice a colonial education. In 1840, after China's defeat in the Opium War, the number of missionary schools increased enormously, and they gradually formed an independent educational system, from primary school to university. Foreign powers directly controlled these schools, with which the Chinese government had nothing to do. In 1890, these schools formed a joint institution known as "China Educational Society" for the purpose of controlling Chinese education. Meanwhile, many of China's feudal warlords and bureaucrat-compradors, who wished to use Western technology to maintain the rule of the Qing government, established a number of Westernized schools for the training of translators, interpreters and compradors. In addition to the "Four Books" and the "Five Classics", these schools also taught "Western" arts (military and industrial technology of the West) and foreign languages — courses that were not offered in the old type of schools. With the development of capitalism in China late in the 19th century, a school of bourgeois reformists appeared. To save China from its impending peril and to strengthen it, they advocated the abolition of the imperial examination system, the reform of the outmoded educational system, and the establishment of new schools. A number of new schools were indeed established in various places of the country. During the late 19th and early 20th centuries, the tottering Qing government, under pressure of circumstances, started to conduct educational reforms by changing the traditional academies in various parts of the country into modern schools run by the government, where natural sciences and foreign languages were added to the curricula. In the meantime, it promulgated a new educational system similar to that of Japan during the Meiji Restoration. Complete schooling, from primary school to university, was to last 26 years. The imperial examination was abolished, and individuals had the right to open primary and secondary schools. Through the above measures, the Qing government had indeed adopted the formalities of a bourgeois education, but the foundation of feudal education was barely touched. Schools for children of nobility were kept intact; the declared aim was to teach "loyalty and filial piety"; and the most important subject in education was still Chinese classics and history. Priority was given to the teaching of Confucian ethics, and women were denied the right to education. Students, upon completion of their study in school, would still receive such scholastic titles as *ju ren* or *jin shi* as if they had passed the traditional civil service examinations. Thus a semi-feudal, semi-colonial educational system had come about in substitution for a feudal variety.

Following the bourgeois revolution in 1911, bourgeois democrats made important changes of the educational system of the late Qing. They did away with "loyalty to the emperor" as an educational aim, abolished exclusive schools for children of noblemen, and made the period of schooling shorter than that worked out during the late Qing. They abolished all the required courses on Confucian classics, and strengthened the requirements for applied science and professional education. Individuals were given greater freedom to establish schools of their own with the exception of normal colleges. Women acquired the right to education, though limited. In primary schools, girls could study side by side with boys, but they would have to study in their own vocational and secondary normal schools. All this reform did indeed have a positive effect on the development of the Chinese education. However, owing to wars and political chaos during the period when China was ruled by the Beiyang warlords (1912-27), none of the bourgeois reforms was thoroughly carried out.

During the more than 20 years under the Kuomintang rule, the United States strengthened its influence and control over Chinese education. Then there were three types of schools: in addition to those run by the Kuomintang government, there were private schools established by individuals or foreign countries. There were 21 institutions of higher learning, 514 secondary schools and 1,133 primary schools funded by foreign countries, of which the United States was the most important. The length of schooling, as promulgated in 1922, followed the pattern as practised in the United States: 6 years in primary education (4 years in lower primary and 2 years in higher primary school), 6 years in secondary education (3 years in junior middle school and 3 years in senior middle school), and 4-6 years in higher education. In addition to the fascist practice of political indoctrination, there were also in school secret organizations to spy on the activities of progressive teachers and students. Reactionary thought was instilled into the minds of students who were required to study old classics. Education in China was thus kept in a backward state. Shortly before Liberation, 80 per cent of the population in China were illiterate, and the percentage was more than 95 per cent for people in the rural areas. Altogether only 20 per cent of the school-age children were in school, most of them children from rich families. Then the number of schools in all of China were only 300,000, with a total enrollment of 25,000,000, only 5.6 per cent of the whole Chinese population. Besides, institutions of higher learning and secondary specialized schools were mostly concentrated in large cities, such as Beijing, Tianjin and Shanghai, and in coastal regions, while education in border provinces or in remote areas peopled by minority nationalities was extremely backward. In some counties there was not even one middle school.

In sharp contrast with the situation described above, education in the liberated areas, which served the cause of revolutionary war and national liberation, underwent a vigorous development under the leadership of the Chinese Communist Party. First, an educational campaign to wipe out illiteracy through night classes and winter schools was conducted on a large scale among workers and peasants. Second, general education was developed in various forms by both the government and the people, often with subsidies from the government. There were full-time schools, half-day schools, morning, noon, and evening classes as well as classes by rotation. Favourable conditions were thus created for the admittance of children from poor and working families. In the years from 1939 to 1941, enrollment at schools in the Shaanxi-Gansu-Ningxia Border Region accounted for one-fourth of the total number of school-age children. Third, with the establishment of various kinds of cadres' and specialized schools, large numbers of military, political and cultural personnel were trained. Among these schools were the Anti-Japanese Military and Political Academy, the Lu Xun Academy of Arts, the Yanan University, the College for Ethnic Minorities, and the Marxist-Leninist Institute, as well as normal, health and telecommunication schools all of which were established in the Shaanxi-Gansu-Ningxia Border Region during the anti-Japanese war (1937-45). During the period of the Liberation War (1945-49), other schools came into existence, such as the Northeast Military and Political University, the North China University and the Beifang University. The length of study varied, from a few months to a year. In order to train specialists who were needed by the New China in the late period of the Liberation War, a number of specialized institutions of higher learning were established in the liberated areas in north and northeast China. In the universities, new departments of engineering, medical science, foreign languages, business and economics were added. Education in the liberated areas was developed under the abnormal situation of war, and there was a lack of uniformity in the length of study. Then large numbers of trained personnel were urgently needed by the revolutionary cause; by integrating theory with practice, education was made available to workers and peasants in a variety of forms. The experience proved to be very useful when the educational system was restructured after the founding of the People's Republic.

2. DEVELOPMENT OF EDUCATION IN NEW CHINA

After the founding of the People's Republic, the People's Government made fundamental changes in the old

educational system. Schools of all levels and types underwent speedy development. In 1980, more than 90 per
cent of school-age children were enrolled as students.
There were 146,269,600 pupils in primary schools,
55,080,800 students in ordinary secondary schools, 453,600
in agricultural and vocational middle schools, 1,243,400
in secondary specialized schools, 1,143,700 as undergraduate students in colleges and universities, 17,700 as postgraduate students, 11,507,700 children in kindergartens,
and 33,100 students in schools for deaf-mutes. Altogether
215,749,600 students, or 20 per cent of the country's population, were attending schools.

During the past 30 years, large numbers of people
have been trained by schools of all levels and types. Altogether 3,177,600 students have graduated from institutions of higher learning, 5,798,000 students from secondary specialized schools, 52,577,000 from senior middle
schools, 157,108,000 from junior middle schools and
658,000 from agricultural and vocational middle schools.

Much has also been achieved in education for
minority nationalities. For many minority nationalities
before Liberation, there was hardly any school at all.
Keeping records by notching wood or tying knots, some
minority nationalities did not even have written languages
of their own. Today the more than 50 minority nationalities have not only primary and secondary schools but
also their own college students. In 1978, total enrollment
of minority nationalities in schools of all levels amounted
to 10,248,000, 10.3 times that of 1951. Of the total enrollment, 36,000 were college students, 27.7 times that
of 1951, a rate of increase far greater than the rate of
increase for the nation as a whole.

	Number of Schools			Number of Students (in ten thousand persons)		
	1949	1980	Increase of 1980 over 1949	1949	1980	Increase of 1980 over 1949
Institutions of Higher Learning	205	675	3.3 times	11.7	114.37	9.8 times
Secondary Specialized Schools	1,171	3,069	2.6 times	22.9	124.34	5.4 times
Ordinary Middle Schools	4,045	118,377	29 times	103	5,508.08	53.5 times
Primary Schools	346,800	917,316	2.6 times	2,439	14,626.96	6 times
Kindergartens	1,300	170,419	131 times	1.3	1,150.77	88.5 times

Note: The figure for kindergarten enrollment was that of 1946, the highest figure of kindergarten enrollment before Liberation.

	College Students	Secondary Specialized School Students	Ordinary Middle School Students	Primary School Students	Total
1951	1,300	5,000	40,000	943,000	989,300
1978	36,000	59,000	2,467,000	7,686,000	10,248,000
Increase of 1978 over 1951	27.7 times	10.8 times	61.2 times	8.14 times	10.3 times

In old China, women did not have much chance to attend schools. Now not only their enrollment in schools of all levels and types has increased, but their proportion in the total enrollment has also increased.

New China has achieved much in developing adult education as a sharp contrast to what old China failed to do. An educational system for part-time study, which was nonexistent in the past, has been established, and the system begins with elementary schools for the elimination of illiteracy and ends with colleges and universities. Between 1949 and 1979, there was a total of 1,190,000 students graduated from part-time institutions of higher learning, 3,700,000 from part-time middle schools and secondary specialized schools and 32,000,000 from part-time primary schools, in addition to 132,000,000 who had become literate.

The number of teachers in schools of all levels and types has continued to grow. In 1980, the number of professional teachers totalled 9,333,300, 10 times that of 1949, as is shown in the table on p. 14.

	Female Students in Primary Schools		Female Students in Ordinary Middle Schools		Female Students in Secondary Specialized Schools		Female Students in Institutions of Higher Learning	
	Absolute Figure (in ten thousand persons)	Proportion (%)	Absolute Figure (in ten thousand persons)	Proportion (%)	Absolute Figure (in ten thousand persons)	Proportion (%)	Absolute Figure (in ten thousand persons)	Proportion (%)
1951	1,206.3	28	40.1	25.6	10.9	28.5	2.3	19.8
1978	6,570.4	44.9	2,715.5	41.5	29.4	33.1	18.1	24.1

	1980	1949	Increase of 1980 over 1949
Kindergartens	410,700	2,000	205 times
Primary Schools	5,499,400	836,000	6.6 times
Ordinary Middle Schools	3,019,700	67,000	45 times
Secondary Specialized Schools	128,600	16,000	8 times
Institutions of Higher Learning	246,900	16,000	15.4 times

Generally speaking, the development of education in New China has undergone the following four stages:

Successful Completion of Socialist Transformation of Education (1949-57) After the country had been liberated, the Chinese Government first took over all the public and private schools from the Kuomintang, and then recovered the sovereign right of running education by reforming or incorporating all the missionary schools that had been subsidized by foreign countries. Furthermore, it put to an end all the fascist practices that had been carried out by the Kuomintang government in running schools. Drawing upon the experience gained from doing educational work in the liberated areas, it adopted a policy of making schools accessible to workers and peasants. It created favourable conditions for the children of workers and peasants to attend schools by providing scholarships and subsidies. Universities not only were tuition-free but also provided free medical care. By 1957, there had been a noticeable increase of students from families of workers and peasants in schools of all levels. To raise the educational level of cadres who came from worker and peasant families, secondary schools with accelerated courses and continuation schools for worker-and-peasant cadres were established.

The People's Government also paid great attention to the elimination of illiteracy. It made relevant decisions and formed a nationwide association for the purpose of eliminating illiteracy. It compiled and published primers for illiterate peasants and popularized a quick method of learning Chinese characters. As a result, the anti-illiteracy campaign was crowned with success. By

1955, 30 million of the 100 million young peasants had become literate.

After 1952, in view of the need in national construction, the government made the necessary adjustments with regard to the universities and their departments. Comprehensive universities formerly comprised of the liberal arts, science, engineering, agriculture and medicine were split to form comprehensive universities comprised merely of the liberal arts and science and independent colleges and institutes specializing in engineering, agriculture, medicine, normal education, business administration and economics, political science and law, the arts and physical education. The students, as well as the number of institutes and colleges specializing in engineering, medicine, agriculture and normal education, were proportionally increased. By moving old educational establishments to the hinterland where new ones were also added, it was hoped that a better geographical distribution of the institutions of higher learning could be attained. In 1953 after the country had launched the First Five-Year Plan, new students were enrolled and graduates received work assignments under a unified plan. Educational development was channelled into the orbit of the state plan.

During the early period of the People's Republic, the Soviet Union was a model for conducting educational reforms. Given the historical circumstances at that time, there were positive results in terms of teaching materials and teaching methods. There were also mistakes, such as the indiscriminate transplanting of the Soviet practice without reference to Chinese reality, the separation of science from engineering, the inflexible and minute divi-

Students in Colleges of Engineering, Agriculture, Forestry, Medicine and Normal Education (in Ten Thousand Persons)

	Engi-neering	Agricul-ture	Forestry	Medi-cine	Normal Education
1949	3.03	0.98	0.054	1.52	1.2
1957	16.3	3.3	0.6	4.91	11.47
Increase	5.4 times	3.4 times	12 times	3 times	9.5 times

sion of subjects, and the abolition of English in the curricula.

Establishment of a Socialist Educational System (1958-66) For eight years progress was made in education, and there was a large increase in the number of college, secondary and primary school students. In 1957, 904,300 students attended senior middle schools, 4 times that of 1949; 5,376,800 attended junior middle schools, 6.4 times that of 1949; and 64,283,000 attended primary schools, 2.6 times that of 1949. Still there was a shortage of schools to accommodate graduates from junior middle and primary schools. In 1957, the number of graduates from primary schools were 4,980,000, but only 44.2 per cent of them, or 2,196,000, went on to junior middle schools. In the same year, the number of graduates from junior middle schools were 1,112,000, but only 39.8 per cent of them, or 444,000, went on to senior middle schools. In 1957, the number of college students was four times

that of 1949, but there was still 42 per cent of the graduates from senior middle schools who could not find a
place in colleges.   In 1958, the Ministry of Education
began to explore new ways to develop education in areas
where the population was large and funds were less than
adequate. Various programmes appeared, such as work-
study programme in which factories were run by schools,
and part-work and part-study programme. Agricultural
middle schools and vocational and technical schools were
also established. Meanwhile, factories, mines and people's communes used their own financial resources to open
a large number of secondary and primary schools, help-
ing considerably educational development. As teaching
was integrated with productive labour and scientific research, many students undertook graduation projects that
combined the academic activities with production. For
instance, the faculty and students of the water conservancy department at Qinghua University successfully designed the Miyun Reservoir in Beijing. This showed
how the combination actually worked.

In 1958, after having had nine years of experience
in running education, the State Council, in enunciating
its educational policy, stated that "education must serve
proletarian politics and be combined with productive labour" and that "everyone who receives an education"
should be enabled "to develop morally, intellectually and
physically and become a worker with both socialist consciousness and culture". A variety of forms would be
adopted to advance the educational cause including schools
established by the state, factories, mines, enterprises
or people's communes; ordinary education combined with
vocational (technical) training; adult education with preschool education; full-time schools with part-work and

part-study schools; formal schooling with private study (including study in correspondence schools and radio broadcast schools) and education with or without tuition. The cause of education received a great impetus as a result. In the course of implementing these programmes, however, there were many "Leftist" mistakes that resulted from a blind development of all types of schools for the purpose of attaining high numerical figures beyond possibilities. The enlargement of the scope of attack in the anti-Rightist movement in 1957 dampened teachers' enthusiasm for education, and the one-sided stress on education in serving proletarian politics and the integration of education with productive labour led to a deterioration in educational standards, as students and teachers had devoted too much time to production and social and political activities.

Beginning in 1961, the various educational organs, by systematically analyzing the experience and lessons that had been gained after the founding of the People's Republic, reduced the number of schools to be established and readjusted the Party's policy towards intellectuals. During the period from 1961 to 1963, to make sure that full-time schools had the right course to follow, the Ministry of Education formulated regulations applicable to the institutions of higher learning and secondary and primary schools with reference to their task and objectives, curricula, teaching, scientific research, productive labour and ideological education. Having prepared an entire set of teaching programmes and teaching plans, the Ministry made further efforts in rectifying the work of teaching and study in various schools. In addition, it also compiled a number of good textbooks. During this period, the number of textbooks compiled for institutions of higher learning and secondary specialized schools in the fields of science, engineering, agriculture and medical science amounted to 3,000 titles; the number of titles was 65 in the field of the liberal arts. There was also a new set of textbooks for a study period of 12 years in primary and secondary schools. Teaching standards in the country's schools reached an unprecedented high level after the founding of the People's Republic.

In 1964, basing upon the experience gained from 1958 in running part-work and part-study schools, and incorporating this experience as an important part of the educational system, the country's educational authorities made further plans for part-work and part-study education. By 1965 steady progress had been made in developing education of all levels and types.

In 1965, there were as many as 434 institutions of higher learning in China, with a student body of 674,000, respectively 2.1 and 4.3 times the peak performance before Liberation. The number of students in secondary schools was 14,320,000, 7.6 times the peak performance before 1949. Enrollment in primary schools was 116,210,000, or 4.9 times the peak figure before Liberation, and 85 per cent of the school-age children attended school. The development of part-work and part-study education also yielded good results. A total of 177 part-work (farming) and part-study universities was set up on a trial basis by full-time institutions of higher learning, and there were 109 universities, with a student body of 29,000, that were either independent or run on a trial basis by factories and people's communes. Moreover, there was a total of 61,626 part-work (farming) and part-study middle schools and agricultural (vocational) middle schools with a student body of 4,430,000. Still, even a greater achievement was made in spare-time education. Altogether there were 410,000 students in spare-time institutions of higher learning, 8,540,000 students in spare-time secondary schools and 29,600,000 students in spare-time primary schools and literacy classes.

In the 17 years before the "cultural revolution", a total of 16,000 postgraduates and 1,500,000 graduates were trained in China's full-time universities. In addition, there were 200,000 students who had completed their study in spare-time institutions of higher learning by correspondence or other forms of spare-time education; 2,950,000 students graduated from secondary specialized schools and 20,000,000 persons were trained as a reserve labour force by ordinary and agricultural (vocational) secondary schools. Education did its part for China's construction.

Education During the Period of the "Cultural Revolution" (1966-76) In 1966 when the "cultural revolution" began, the two counter-revolutionary cliques, headed by Lin Biao and Jiang Qing respectively, moved first to the cultural and educational field and usurped the power of leadership. In 1971, they convened a national conference on education, where they concocted a so-called "Summary from the National Conference on Education" and altered and distorted the Party's policy on education. Negating the experience and achievement gained in educational work during the 17 years before the "cultural revolution", they imposed a feudal, fascist dictatorship on the educational front, causing unprecedented harm to the country's educational work.

Lin Biao, Jiang Qing and company reviled schools under the leadership of the Communist Party as old-type schools controlled by the bourgeoisie. By maligning the majority of teachers and students as bourgeois intellectuals who had been trained in the old-type schools, they subjected them to ruthless persecution and attack. They abolished all rules and regulations which had been proved effective during the previous 17 years on the ground that they were revisionist, and they put an end to the entrance examination system in higher education, the system for training postgraduates and the working rules for all schools. During the period from 1966 to 1969, there was no new enrollment in all institutions of higher learning and secondary specialized schools; a number of secondary and primary schools suffered the same fate. Altogether 106 institutions of higher learning, plus a large number of secondary specialized schools, were either abolished outright or broken up by merger, moving, or dispersal. Their books and equipment were lost, and their buildings were occupied for other purposes. The part-work and part-study education came to an end; so did agricultural and other secondary vocational schools. After schools were reopened in 1970, Jiang Qing and company played up the theory that workers, peasants and soldiers "should attend, administer and reform the universities". Under the pretext that one must oppose "what the teachers say goes", they preached anarchism to undermine school discipline and stirred up ill feeling between teachers and students. As a result, teaching in all schools became chaotic and teaching standards deteriorated.

In the 10 years between 1966 and 1976, only 1,033,000 students graduated from China's institutions of higher learning. Of them 670,000 were enrolled in 1965. This means that at least 1,000,000 badly needed people could have been trained for the state but were not. In 1976, the country had only 392 institutions of higher learning

— 42 less than the year of 1965 — and a student body of 560,000, 16 per cent less than that of 1965. In 1976, the number of students in secondary schools was 59,050,000. As large numbers of secondary specialized schools were closed and all agricultural and vocational middle schools, plus the part-work and part-study programme, came to an end, disruptions were wrought in the internal structure of secondary school education, thus giving rise to disproportion and abnormality in the development of ordinary middle schools.

Education During the New Period (1976 to the present) In the last few years fundamental changes have taken place on the educational front after stupendous efforts have been made to remedy the chaotic situation created by the Gang of Four and set education on the right course.

The year 1976 saw the smashing of the Gang of Four and the re-establishment of the Party's leadership in various schools. Jiang Qing and Lin Biao were subjected to criticism for their crimes in undermining education. Cadres and intellectuals who had been unjustly, falsely or wrongly accused or punished in the course of the "cultural revolution" and in the political movements of 1957 and 1959 were rehabilitated in line with the Party's new policies towards cadres and intellectuals. The former system of recruiting college students under a unified plan was restored, and the quality of new students was assured. Educational rules and regulations were revised and promulgated; professional meetings were called to devise new teaching programmes and compile new teaching materials. With an enlarged enrollment, the former system for training postgraduates was restored. Key schools at various levels were established; academic titles were restored; and an academic degree system was instituted. Once again the initiative of educational workers for achievement and the enthusiasm of students for study were brought into play. Together they gradually restored order in education and improved teaching standards.

China's education was again set on a healthy road towards development after several years of readjustment and rehabilitation. By 1980, considerable progress had been made in schools of all levels and types as compared with 1976, and improvement was also scored in the internal structure and geographical distribution of secondary schools. Agricultural middle schools were not only restored, but new ones were also established. Sparetime education for workers and peasants reappeared. Education by radio and television made its debut in 1979.

As China has now entered upon a new historical period of development, its major task is now socialist modernization. To attain the new educational goal, the Ministry of Education, in 1980, called a meeting on educational work, during which specific tasks were set, to be completed in three to five years. First, the ideological work of schools must be strengthened so as to build up students' confidence in socialism and foster in them the communist spirit and morality. Second, primary education must be made universal by the end of the 80s. Various measures shall be adopted to keep children in school for a five-year period, so they can attain the actual level of graduation. In areas where conditions permit, secondary education must be made universal too. Third, the secondary education should be so restructured as to reduce the scale of development for senior middle schools where technical and vocational courses should be grad-

Number of Schools of All Levels and Types in 1980 Compared with That in 1976

	Higher Educational Institutions		Secondary Specialized Schools		Ordinary Middle Schools	
	Number of Institutions	Number of Students (in ten thousand persons)	Number of Schools	Number of Students (in ten thousand persons)	Number of Schools	Number of Students (in ten thousand persons)
1980	675	114.37	3,069	124.34	118,377	5,508.08
1976	392	56.5	2,443	69	192,152	5,836.5

	Agricultural (Vocational) Schools		Primary Schools	
	Number of Schools	Number of Students (in ten thousand persons)	Number of Schools	Number of Students (in ten thousand persons)
1980	3,314	45.36	917,316	14,626.96
1976	0	0	1,044,300	5,005.5

ually added and, in the meantime, restore and develop secondary specialized, technical and agricultural middle schools where a labour force of the highest quality is to be trained. Fourth, there must be concentrated efforts in creating a number of key schools, on the primary and secondary as well as college level, so they could serve as true examples for other schools to follow. Fifth, with emphasis on consolidating and raising the quality in higher education in three to five years, efforts should be made to improve the level of teaching and scientific research. When quality is no longer a question, steady steps should be taken to develop fields of study that remain weak or nonexistent. Sixth, scientific and cultural exchanges with foreign countries must be actively developed. Seventh, education for minority nationalities and in the border areas should be strengthened. Eighth, there must be a vigorous development of adult education. Within three to five years illiteracy among young and middle-aged adults must be basically eliminated; young workers in mines and factories should be taught through various channels to attain the level of a junior middle school graduate. New colleges for workers and staff members should be developed; so should education by radio and television. People would be trained as the state needs them to modernize the country, irrespective of their educational background.

3.   PRESENT EDUCATIONAL SYSTEM

Today there are in China three independent school systems: full-time schools, spare-time schools and part-work and part-study schools.

(1) FULL-TIME EDUCATION:

Full-time schools at all levels form the bulk of China's school system and play the main role in the country's education. They are grouped as follows:

Full-time primary schools. Children are enrolled at the age of six or seven. They study for three years in the lower grades and two years in the higher grades, for a combined schooling of five years.

Full-time secondary schools. Within this category are junior and senior middle schools. Upon completion of study for three years in junior middle schools, many graduates will be enrolled, through entrance examination, in various types of senior middle schools. Those who fail to enroll will become industrial or agricultural workers. A two-year study is specified for senior middle schools. Upon completion of study, a small number of graduates would successfully pass the college entrance examination to continue their study at institutions of higher learning, while the majority of them would wait for job assignments. The schooling in agricultural (vocational) middle schools lasts three years.

Secondary specialized schools. Schools of this kind enroll graduates from junior middle schools, and schooling lasts for three or four years — sometimes for as long as five years for some special curricula. The period could be two to three years if an enrollee is a senior middle school graduate. For full-time secondary schools specializing in the performing art, such as dance, music and operas, children at the age of 11 or 12 would be enrolled to study for six to seven years, regardless of whether they have or have not completed their studies at primary schools. The secondary normal schools accept graduates from junior middle schools for a schooling period of three years. Graduates from secondary specialized schools must work for two years before they are allowed to take entrance examination to enter colleges of their own fields.

Students of regular majors in institutions of higher learning generally study for four years; for some majors in engineering and science the length of study could be five years; in medical science some majors require a study period of six years. Individual universities or colleges may even require their students to study for as many as eight years. Vocational colleges generally have a study period of two to three years. As for post-graduate study, the length of training could be anywhere from two to five years.

In areas inhabited by minority nationalities, native tongue is used as a medium of instruction at schools of all levels. One year of study would be added if a school also teaches the Han language.

(2) SPARE-TIME EDUCATION

Since the founding of the People's Republic, special attention has been paid to spare-time education for workers, peasants and cadres. During the educational reform in 1951, spare-time education was formally incorporated as part of the state's educational system. A complete spare-time system that ran from primary schools to colleges were gradually established.

Before 1966, the length of study for spare-time education was as follows: Primary schools, five years; junior middle schools, three to four years; senior middle schools, three to four years; and colleges, three to five years. However, due to complexities involving spare-time study, many schools found it impossible to follow the uniform length of study. In 1979, a change was made in the length of study for spare-time primary and middle schools. A student would be regarded as having completed his study and be granted a diploma if he had earned a fixed number of class hours and then passed a written examination, regardless of his length of stay in school. His record of formal schooling would be recognized by the state.

Literacy class. Workers and others are required to know 2,000 characters and achieve "four ables" — able to read, write, recognize and speak these characters. As for peasants, the requirement is 1,500 characters and the achievement of "four ables". They have to pass the examination before they can be counted among the literates.

Spare-time primary schools. Altogether 360 class hours are required for graduation.

Spare-time secondary schools. They are divided into junior and senior middle schools, each requiring 720 class hours for graduation.

Spare-time institutions of higher learning. The length of study is four to six years, approximating 2,000 to 2,500 class hours.

(3) PART-WORK AND PART-STUDY EDUCATION

Part-work and part-study education began in 1958. However, not until 1964 did the state decide to experiment with it on a nationwide scale. Within the system were part-farming and part-study primary schools, part-work and part-study secondary schools, part-work and part-study secondary technical schools, and part-work and part-study universities. All these schools, however, were abolished during the "cultural revolution". Now an active effort is being made to restore them, as they are recognized as a component of the whole educational system. Still at its infant stage and run on a trial basis, part-work and part-study education lacks uniformity in the length of study.

Since at present the educational system of China is anything but complete, educational authorities are drawing from past experience to conduct necessary reforms. One of the proposed reforms is to lengthen the period of study for senior middle school students so as to assure the quality of their education. In 1980, 14 senior middle schools in Beijing lengthened the period of study from two to three years. Moreover, efforts were made to set up part-work and part-study and spare-time schools of all levels in order to make the educational system more complete.

4. EDUCATION OF ALL LEVELS AND TYPES

(1) PRESCHOOL EDUCATION

Brief Introduction Since the founding of the People's Republic there has been rapid progress in preschool education, even in rural areas. In 1980, China had 170,419 kindergartens with a total enrollment of 11,507,700 children, or 89.5 times that before Liberation. But the number of children enrolled in kindergartens at present

Full-time School System

Grade

Postgraduate Studies Programme

5

4

3 — Specialized
Colleges for
2 — Professional
Training

1

Institutions of
Higher Learning

Agricul-
tural
Middle
Schools

Voca-
tional
Middle
Schools

Secon-
dary
Specialized
Schools

17 1

16 5   Senior
Middle
15 4   Schools

14 3

13 2   Junior Middle Schools

12 1

11 5

10 4

9 3   Primary Schools

8 2

7 1

6

5

4   Kindergartens

3

age

constitutes only a small proportion of the children of the same age. In 1979, to speed up the development of preschool education, governments at various levels formed leadership groups to be responsible for child-care work. Meanwhile, the Ministry of Education promulgated regulations on the work of child care and strengthened the research work on preschool education and the training of kindergarten teachers. At this moment, 28 schools for the training of kindergarten teachers have been reopened. In such provinces as Shandong, Anhui, Sichuan and Hubei the secondary normal schools offer courses for the training of kindergarten teachers, normal colleges add departments of preschool education, and medical schools have special classes for the training of maternity and child-care workers. Many provinces and municipalities have formed research societies to study preschool education. Since 1980, well-equipped primary schools have offered kindergarten classes for preschool children. Meanwhile, government departments, factories and mines did their best in opening their kindergartens to the general public. In 1980, 71 government departments, factories and mines in Beijing opened their kindergartens to local citizens not related to them.

A kindergarten could be run by either the state or civilians. In the former category are those run by governmental agencies, educational departments, army units and state enterprises. In the latter category are those run by neighbourhoods in the cities and people's communes and production brigades in the rural areas.

The Task of Kindergartens   A kindergarten provides children with necessary nourishment and cultivates in them good habits to ensure them a healthy development in mind as well as in body. It gives them good guidance regarding a gradual understanding of their social and natural environment. It helps them develop their intellect and their ability in verbal expression. It teaches them simple calculation and cultivates in them the interest in study. It assists them in acquiring fine moral qualities: honesty, bravery, cooperation, friendliness, discipline and politeness. It teaches them some elementary knowledge and skill on music, the fine arts and dance and cultivates in them love for art.

Organization of Kindergartens There are overnight, day, and mixed kindergartens. Many primary schools have preschool classes under their jurisdiction. In the countryside there are seasonal classes opened for preschool children during busy seasons.

Kindergarten children are grouped according to their age: 3- to 4-year-olds belong to the lower class, 4- to 5-year-olds belong to the middle class and 5- to 6-year-olds belong to the high class. Each class has 20 to 35 children. Overnight kindergartens have a slightly smaller number of children for each class.

Within each kindergarten there is usually a nursery to take care of 2- to 3-year-old toddlers. To help working couples, many nurseries accept babies that are 56 days old or older.

Game and Homework The education of kindergarten children takes the form of game playing and homework in various forms.

Game playing is the children's main activities. These are creative games, such as role playing, house building, and theatrical performances. There are also games of sports, intellect, and music. Each day two periods are

set aside for activities of this kind, and each period lasts
one to one and a half hours.

The kindergarten teaches such subjects as language,
general knowledge on life and natural environment that
could be easily understood, mathematics, music, fine arts
and physical culture. The number of class hours each
week vary according to the age of the children: six periods
for the lower class, each lasting 15 minutes; 12 periods
for the middle class, each lasting 25 minutes; and 14
periods for the high class, each lasting 35 minutes.

Hygiene and Health Protection All kindergartens
are concerned with the scientific arrangement of the
children's life. The children must have 11 to 13 hours of
sleep daily (including noonday rest) and sufficient outdoor
activities. There are full-time or part-time medical and
child-care personnel responsible for health protection,
and the children are given preventive inoculation when
needed. Local hospitals and clinics give the children a
physical examination once every year and treat such
abnormalities as trachoma, weak sight, slant eyes and
tooth decay. They have on file the health record of every
child.

(2) PRIMARY EDUCATION

Brief Introduction Full-time primary schools have
as their task the training of a new generation and lay a
good foundation for the children in preparation for future
education in middle schools.

Since the establishment of the People's Republic, the
state has consistently emphasized the importance of
education. Over the last few years local authorities have

focused the popularization of primary education as a main task in educational development in the rural areas. In mountainous areas with poor economic conditions or border regions inhabited by minority nationalities, various forms of education have been adopted in running schools. As local problems are solved one by one, the attendance rate of school-age children has been increasing year by year. In the remote and inaccessible mountainous areas where inhabitants are scattered, primary schools of the simplest form have been established. To accommodate children who cannot attend class in regular hours, some primary schools in the countryside even have morning, noon and evening classes. Moreover, attention has also been paid to the enrollment of over-age children. In 1980, 93 per cent of school-age children actually attended school. But the progress varied from place to place. Many pupils, after entering school, could not complete their studies. Various measures will have to be taken in the future not only to increase the enrollment of school-age children but also to make sure that they will complete a five-year study and thus graduate.

To raise the teaching standards in primary schools, the Ministry of Education in 1978 decided to establish a number of key schools as models for others to follow. There are now in China about 7,000 key primary schools.

The Objective of Primary Education and Curricula
The purpose of primary education is to cultivate in the pupils the moral character of loving the motherland, the people, physical labour, science and public property, cherishing socialism and supporting the Communist Party; to help them to acquire the ability of reading, writing and arithmetic, fundamental knowledge in natural and social sciences, and good study habits; and

to enable them to enjoy good health and acquire fine habits in life and labour.

Full-time primary schools have a study period of five years. Each year there are 40 weeks for study (including four weeks for reviews and examination), 10 weeks for summer and winter vacations, and two weeks for miscellaneous activities.

Curricula include ethics, Chinese, mathematics, natural science, foreign languages (offered by schools with favourable conditions), geography, history, physical culture, music, fine arts and physical labour (for students in the fourth and fifth grades).

There are no manual labour requirements for pupils in the first, second and third grades; they are asked only to do such work as they are able to according to their age. Pupils in the fourth and fifth grades have one class hour of manual labour per week, taking part in some social labour activities that are organized by school authorities, such as planting trees and cleaning public places. Some schools even have small factories of their own. In busy seasons, pupils in the countryside provide some supplementary labour service in production teams or at home.

There is a total of 4,644 class hours in the five years of primary education. A weekly number of 24-27 class hours is scheduled for each grade. Actually, there is no uniformity in this matter, as differences exist in various places.

For Chinese and mathematics, a written examination is held in the middle and at the end of the semester. For all other courses, a written examination is held at the end of the semester. Pupils who fail in make-up examinations for Chinese or mathematics would not be allowed to graduate or to be promoted to a higher grade. Pupils who do exceptionally well in the examinations would be allowed to skip grades.

Life and Study   Each day, pupils at primary school generally have less than six hours for class meetings, private study, manual labour and extracurricular activities. For after-class activities on a weekly basis, there are two class hours for private study, two class hours for scientific, technical and recreational activities, and one class hour for weekly meeting and Young Pioneers activities. The teachers are required to instruct their pupils to do well with their homework. To ensure them to have 10 hours' sleep, they must restrict the amount of homework.

(3) SECONDARY EDUCATION

Brief Introduction   In line with the principle of all-round moral, intellectual and physical development, the task for an ordinary middle school is to train socialist-conscious and cultured workers, namely, qualified skilled workers for society and qualified students for institutions of higher learning.

During the "cultural revolution", there was a sharp increase in the number of full-time ordinary middle schools. A lopsided development of secondary school education came as a result following the elimination of agricultural and vocational middle schools and a reduction in the number of secondary specialized schools.

At present only about 40 per cent of junior middle school graduates and about 4 per cent of senior middle school graduates are able to continue study, while the

majority of the graduates have to take part in productive
labour. Owing to the lack of special training, the majority of the graduates have a difficult time to find
suitable work. In view of this problem, middle schools,
beginning in 1978, have carried out necessary reforms.
In 1979, 18,100 ordinary middle schools were closed,
resulting in a decrease of enrollment by 6,433,000.
Meanwhile, 2,107 agricultural middle schools, with a student body of 227,000, were restored or established. In
addition, 31 secondary vocational schools, with an enrollment of 8,400, were reopened or established. In the
future there will be more ordinary senior middle schools
changed into agricultural (vocational) middle schools.
Ordinary middle schools will offer more technical and
vocational courses to prepare students well for future
work. In 1980, the city of Beijing, on a trial basis, opened
more than 90 vocational senior middle school classes with
an enrollment of 4,000 junior middle school graduates.
Among the curricula are 35 subjects such as radio, television, electric automation, machine building, tourism,
foreign trade, catering trade, cookery, accounting, printing and construction. The length of study is three years.

In 1979, more than 5,700 middle schools were desig-
nated as key schools in various provinces, centrally
administered municipalities and autonomous regions.
They have better teachers and equipment.

The Purpose of Training The purpose of middle
school education is to cultivate in the students patriotism
and communist ethics; teach them gradually to foster
the proletarian world outlook and outlook on life and
work for the people and the modernization of the country; help them do a good job in studying and mastering
the elementary knowledge of science and technology and

culture; and ensure them a normal development in body
and mind and enable them to acquire aesthetic judgement
and labour skills.

Curriculum and Teaching Plan Gradually, as far as
full-time middle schools are concerned, the length of
study has been changing from five to six years. For each
academic year 40 weeks are devoted to academic learning,
leaving 10 to 11 weeks for winter and summer vacations,
holidays and festivals. In addition, four weeks are set
aside for physical labour and technical training. Middle
schools in the countryside allow students to take time
off for agricultural work as a substitute for physical
labour requirement in school and schedule vacations
according to the change of seasons. The students generally have at least one month of rest per year.

The courses offered in junior middle schools include
Chinese, mathematics, foreign language, politics, history,
geography, biology, physics, chemistry, physiology,
physical culture, music, and fine arts. The courses offered in senior middle schools include Chinese, mathematics, foreign language, politics, physics, chemistry,
biology, history, geography and physical culture.

At present many schools have tried to offer courses
in labour skills (including technical and vocational
courses) in senior grades. In key middle schools, students, beginning in the second year of the senior class,
can take four class hours of elective courses. Many schools
separate courses of liberal arts from those of science in
teaching.

For middle schools of the five-year kind, there is
a total of 4,898 class hours for academic learning. For
those of the six-year kind, the number of class hours
is 5,554. In junior middle schools there are 30 to 31 class

hours each week for academic learning. The number is
26 to 29 class hours for senior middle schools.

Training in Labour Skills The main purpose is to
cultivate in students the correct attitude towards labour
and to provide them with the opportunity to acquire
labour skills. Labour skills include skills in industrial
and agricultural production, basic skills in service work,
vocational skills and skills in doing work for the public.
Students work either in school-run factories or farms,
in regular factories or, by contacting people's communes,
in production teams in the countryside. While engaging
in productive labour, students also make visits to the
farms and workshops. Workers and peasants are request-
ed to give them lectures on ideology. School-run factories
and farms, where labour is provided by teachers and students, do not pay income tax to the state. Instead, the
income would be used to subsidize teachers' and students'
food, clothing, and transportation cost or to advance their
general welfare. It could also be used to buy school
equipment.

Extracurricular Activities and Life Regulation In
full-time middle schools commuter students form the
majority of the student body. Only in a few key schools
are there a number of boarding students. The time for
students actually at school does not exceed seven hours
per day: five to six hours for classroom learning and
private study and one to two hours for extracurricular
activities. Extracurricular activities in middle schools are
organized and arranged by the dean's offices, the form
masters, the students' unions, the Youth League and the
Young Pioneers. In schools are various literary, art,
athletic and science organizations or societies under the
guidance of teachers appointed by school authorities who

also provide funds and equipment. Invariably these organizations or societies are very active, conducive to the development of students' body and mind. To ensure the students' healthy growth, school authorities require them to have one hour's physical exercise and nine hours' sleep each day in junior middle schools and eight hours' rest in senior middle schools. The adolescents are also taught physiology and personal hygiene. Students are not allowed to drink wine or smoke cigarettes; they are persuaded not to fall in love, so they can concentrate their energy on study. As for those students who have financial difficulties, they are either given subsidies by the state or are partially or totally exempted from tuition payment.

Recruitment and Work Assignment Different middle schools have different practices in recruitment. For the entrance examination in 1980, the questions were the same for all the districts and counties in Beijing, but each district or county graded its own papers. Students took the examination in sites within walking distance, and schools would choose those who had the best scores. For primary school graduates, the graduation examination was the same as the entrance examination to junior middle school. The day of the examination and the questions in the examination were the same throughout the city. Those who failed the examination would have to remain in junior middle school or primary school to continue the study.

The government's labour departments and the work-assignment offices for educated youth in various residential districts of a city are in charge of work assignments for middle school graduates. Many governmental enterprises and institutions are, on an experimental basis, using written examinations as a means to evaluate prospective employees who are tested not only on their knowledge of various subjects but also on their health and ethics. Only those with best performance will be selected. Their schools can offer opinions of their own regarding the candidates' academic achievement and ideological quality.

Agricultural Middle Schools The new or reopened agricultural middle schools are on the whole senior agricultural middle schools. There are, however, some junior agricultural middle schools. Schools of both kinds are controlled either by the state or by the people's communes or production brigades with subsidies from the state. The main task of these schools is to train reserve forces of labour for the people's communes and production brigades. In addition to the courses offered in ordinary middle schools, the students must also study agricultural science and learn some skills of production. The length of study for these schools is two to three years. After graduation, most of them would return to the people's communes and production brigades to do their part in modernizing agriculture. Some of them may take entrance examination and enter institutions of higher learning.

(4) SECONDARY SPECIALIZED EDUCATION

Task and Present Situation The purpose of secondary specialized schools is to train specialized personnel with political awareness and professional proficiency. The end product, being morally, intellectually and physically fit, will ably contribute to the cause of socialist modernization. The schools shall see to it that the students will have communist morality, support the leadership of the Communist Party, love socialism, and are determined to serve the socialist cause and the people. At the same time they shall attain the level of a senior middle school education. They shall master the basic theories and specialized knowledge of their own fields and have practical skills.

Over the last one year or so, there has been a speedy development in secondary education. Today there are in the country 3,069 secondary specialized schools, of which 239 are key schools. Total enrollment is 1,243,400, an increase of 39.8 per cent over that of 1978. Total number of teachers and staff is 298,400, of which 128,600 are full-time teachers, 61,300 and 29,000 respectively more than in 1978.

Fields of Study In the category of secondary specialized schools are secondary technical schools and secondary normal schools. There are as many as eight different types of schools in the category of secondary technical schools: engineering, medicine, agriculture, forestry, business administration, political science and law, physical culture and art, comprising altogether 345 subjects. Among these, 242 are related to engineering, 25 to agriculture, 11 to forestry, 12 to medicine, 34 to business administration, and 20 to art.

Length of Study There are two kinds of secondary specialized schools in terms of length of study. Those schools that accept graduates from junior middle schools as freshmen have a four-year course of study as a rule. The length of study could be five years in some particular cases. Some schools, however, choose to keep the old three-year system. Those schools that accept graduates from senior middle schools as freshmen have generally a two-year course of study. For certain subjects related to medicine and engineering, the studying period may be two and a half or three years. Secondary art schools that accept 12-year-old graduates from primary schools as first-year students to study dance, music and drama, etc., may have a six-year course of study. Some schools specializing in the theatre may take 11-year-olds from primary schools and require them to study seven years before they are allowed to graduate.

Curricula and Evaluation of Study Courses offered by secondary specialized schools fall into four major categories: politics, general knowledge, basic professional skill and specialized skill. About 75 per cent of class hours are used for the teaching of the first three while the latter takes up 25 per cent. The number of courses varies from one discipline to another; engineering, agriculture, forestry, and medicine each has about 20 courses.

There are approximately 3,500 class hours in teaching for schools that have a four-year period of study, 2,600 class hours for schools that have a three-year period of study and 2,000 class hours for schools that have a two-year period of study. During each academic year two months are set aside for winter and summer vacations.

To strengthen the students' technical training, secondary specialized schools offer practice courses. Engineering and agricultural schools generally have factories and farms of their own; schools specializing in business administration and other subjects keep close contact with agencies and departments where students can practise what they have learned. Factories and farms attached to schools are tools of instruction, not those of profit, though they do produce. The ratio between academic learning and practice is 7 to 3 for such disciplines as engineering, agriculture, forestry and medicine, and 8 to 2 for business administration.

The students' academic performance and moral character are evaluated at the end of each term and each academic year. Diplomas will be granted to those who have successfully completed the required academic and practice courses and graduation projects, and have, in addition, passed the required examinations. Diplomas can also be granted to those who fail in the first but pass the make-up examination. As for those who should fail the second time, they can only be given certificates showing the courses completed. Upon their application, make-up examination will be given within one year after they begin to work, and diplomas will be issued should they pass the examination.

Generally speaking, students in secondary specialized schools are resident. The tuition is free, and 75 per cent of them are given subsidies by the state. Majors in coal mining, art, physical culture, nursing, midwifery and teaching are completely subsidized by the state.

Recruitment and Work Assignment Generally speaking, secondary specialized schools recruit students from provinces where the schools are located. Upon graduation the students are assigned work by the provincial authorities under a unified plan. Schools of some fields of study recruit students in provinces other than their own, but the students, after graduation, must return to their native places for work assignments. Most graduates from secondary specialized schools are assigned to work in enterprises or institutions as technical cadres. Some, however, would be employed as skilled workers doing complicated work on an advanced level.

(5) HIGHER EDUCATION

Brief Introduction In 1980, there were in the country 675 institutions of higher learning, and there was plan to establish 15 more. Total number of students accepted as freshmen was 281,200. In the same year, senior middle school graduates numbered 6,161,500, and only 4.6 per cent of them went to college. In 1978, the proportion of college students among Chinese population was 8.9 per ten thousand. In 1980, institutions of higher learning had a total of 246,900 full-time teachers, or one teacher for every 4.6 students. Though there has been speedy development in higher education over the last few years, because of existing conditions, it cannot yet meet the need of the country and satisfy the demand of young people for further education. In 1978 some colleges began to accept commuter students who numbered more than 100,000. Because of this success, continued efforts will be made in the enrollment of commuter students. In addition, there will be more evening and correspondence colleges.

Over the years, colleges of science and engineering have been expanded to the detriment of liberal arts, law and business administration. According to the 1978 statistics, the percentages of students majoring in different subjects are as follows: engineering, 33.6 per cent; science, 7.5 per cent; medical science, 13.2 per cent; agriculture, 6.3 per cent; forestry, 0.9 per cent; teacher-training, 29.2 per cent; liberal arts, 5.4 per cent; business administration and economics, 2.1 per cent; political science and law, 0.2 per cent; physical culture, 1 per cent; and art, 0.6 per cent. As there is still a shortage of personnel for business administration and law, efforts must be made in the development of these disciplines.

In 1980, there were 97 key colleges and universities throughout the country.

Aim of Training in Institutions of Higher Learning Institutions of higher learning have as their task the training of highly specialized personnel and the development of science and technology. They aim to produce students who are socialist-minded, sound in body and spirit, well-versed in the theories, knowledge and skills of their own fields, and aware of the present state, as well as the past, of their own branches of sciences. In addition, they must know well one foreign language so they can read professional books and magazines published in that language.

Classification and Length of Study There are 12 kinds of institutions of higher learning: comprehensive universities, colleges and universities of engineering, agriculture, forestry, medical science, teachers' training, business administration and economics, political science and law, physical culture, arts, foreign languages, and minority nationalities.

A comprehensive university has two colleges: liberal arts and sciences. A college of liberal arts has, generally speaking, the following departments: languages, literature, history, philosophy, economics, law, and others. A college of sciences has, generally speaking, the departments of mathematics, physics, chemistry, biology, geography, geology, radio, electronics, and others. Beijing, Nankai and Fudan universities are examples of comprehensive universities.

There are two kinds of engineering colleges or universities. One is the multi-discipline engineering college or university that has many engineering departments besides the school of sciences. Qinghua University is a good example of this kind. It has such departments as construction engineering, water conservancy, mechanical engineering, precision equipment, thermal energy, electric power, automation, engineering physics, engineering chemistry, computer engineering, radio and electronics. The other is the single-discipline engineering college such as the colleges of mining, iron and steel, geology, petroleum and chemical engineering.

Subjects of Study Institutions of higher learning offer academic programmes in accordance with the need of the state for construction and the development of sciences. Since the founding of the People's Republic, an increasing number of academic disciplines and subjects has been offered, especially in the fields of sciences and engineering. In 1978, Chinese institutions of higher learning had in them more than 800 separate disciplines, of which 500, or 64 per cent, were those of science and engineering.

Courses Offered Courses offered at institutions of higher learning would be required, elective, or special lectures.

Among the required subjects are a): such interdepartment courses as foreign languages, political theory (history of the Communist Party, political economy, dialectical materialism and historical materialism, and history of international communist movement), physical culture (for freshmen and sophomores); b): basic courses (including basic courses on the students' majors) that are offered in the freshman and sophomore years; and c): specialized courses taken during junior and senior years.

Elective courses are offered in accordance with the need of the subjects and the qualifications of the teaching staff. They can be taken by non-majors.

Special lectures are given irregularly and the attendance is optional. They are meant to introduce the recent achievements or developments of various fields of study.

Scientific Research Institutions of higher learning undertake scientific research as one of their important programmes. Many of them have established institutes specially for this purpose. For instance, Beijing University has 11 research institutes, including those of mathematics and theoretical physics, Qinghua University has a research institute of nuclear energy, Nankai University has a research institute of economics, and Liaoning University has a research institute of Japanese studies. All of them are well known in the country.

Research institutes in institutions of higher learning are responsible for research tasks as assigned by the state. They are commended by the state in accordance with their outstanding and verified achievements. Statistics show that of the 108 key projects of scientific research certified by the state in 1978, institutions of higher learning took part in 70 per cent of them. Of the 70 per cent, one-third had been initiated and organized by them. The 30 universities under the direct jurisdiction of the Ministry of Education were responsible for the achievements made in 1978 in the more than 400 key items of research in such fields as mathematics, physics, chemistry, biology, geology, geography, mechanical engineering, civil engineering, chemical engineering, radio and electronics.

Institutions of higher learning received more than one hundred commendations from the National Science Conference held in 1978. Nankai University received 25 and Beijing Iron and Steel Institute 24. Particularly outstanding in the research results were the synthetic crystalline bovine insulin created jointly by Beijing University and the Chinese Academy of Sciences, the theory of granite formation by Nanjing University, etc.

Scientific research in the institutions of higher learning is undertaken by scientific research personnel, teachers and postgraduates. Teachers are generally given one-third of their time to do research.

Productive Labour and Social Investigation (Production Practice) Generally speaking, college students must participate in industrial and agricultural production and perform physical labour in the interest of the public for a total of about 10 weeks. In addition, they must participate in social investigation or production practice for a total of about 12 weeks.

Most institutions of higher learning either have small factories of their own or keep in close touch with outside factories or farms and use them as training grounds for students to do productive labour. There are two types of school-run factories in the institutions of higher learning. One type serves as means of teaching and scientific research, and making profit is only second in importance. The other type, while related to certain academic disciplines on the campus, carries out production in accordance with state or local plans and has its own independent system of accounting.

Training of Postgraduates The main purpose of postgraduate studies is to train high-quality specialized personnel for teaching or research in the institutions of higher learning. The postgraduates are required to be well grounded in Marxism-Leninism, have solid and systematic knowledge of their respective fields, and be competent to do scientific research. In addition, they must know two foreign languages, in one of which they must be proficient. They must be capable of doing independent research and also of teaching.

In 1980, the country had a total of 21,600 postgraduates in 316 institutions of higher learning and 160 other institutions, including the Chinese Academy of Sciences and the Chinese Academy of Social Sciences. Most postgraduates studied engineering, followed by science and medicine.

The length of study for postgraduates varies from two to four years. As for in-service postgraduates, the length of study is fixed at three to five years.

Experts and faculty at large are jointly responsible for the training of postgraduates. Research groups, headed by renowned experts, are in charge of guidance. In some disciplines the tutorial system is adopted. A postgraduate divides his time equally between taking courses and doing research. His graduation thesis must be defended successfully before an evaluation committee and approved by the president of the university or college before he is allowed to graduate.

Recruitment and Work Assignment Students for the country's institutions of higher learning are recruited under a unified plan. Under the guidance of educational authorities at different levels, recruitment committees are formed to recruit students either on a nationwide scale or locally.

Before taking entrance examinations, prospective students are given an introduction on the subjects of study available in the institutions of higher learning. As the purpose of education must be in line with the need of the state, they must consider the state interest as well as their personal preference when deciding the correct course to take.

As all administrative work, including the evaluation of examination papers, is in the hands of responsible persons at various provinces, municipalities and autonomous regions, the state is merely responsible for making questions to be used across the country. Taking into account the prospective students' test results, health conditions, personal choice, colleges and universities will elect those regarded as most outstanding. Key institutions of higher learning will have the priority in choosing the best among the successful candidates.

The Ministry of Education is also responsible for the administrative work in enrolling postgraduates. It does this before college graduates receive their work assignment.

The state assigns work to college graduates and postgraduates according to a unified plan. Upon completion of their study, in-service postgraduates generally return to their original units.

Treatment of College Students and Postgraduates
Students in China's institutions of higher learning enjoy free tuition, free medical treatment, and free boarding. They are only asked to make some contribution to food and teaching material expenses. Students with financial difficulties will receive additional help, including food expenses. Students at normal colleges or universities, physical culture institutes and colleges of minority nationalities do not have to pay anything at all. Students studying such subjects as aquaculture, oceanography, navigation,

dance, theatre, and music have to pay only 60 per cent of their food expenses. Students who have worked for more than five years prior to their enrollment in colleges will continue to receive pay from their former working units. The state provides no direct aid in that case.

In terms of financial aid, there are two kinds of postgraduates. Cadre or worker students will continue to receive their formal salaries or wages; while others will enjoy 90 per cent of the salaries normally accorded to newly employed college graduates.

Granting of Academic Degrees Different academic degrees supposedly show different levels of achievement in the institutions of higher learning. Since the founding of the People's Republic, the Ministry of Education has twice formulated regulations governing the granting of academic degrees. Though none of the regulations has been carried out, they lay the foundation for the granting of academic degrees. In 1980, the Standing Committee of the Fifth National People's Congress at its 13th session approved the Regulations of the People's Republic of China on the Granting of Academic Degrees, which was put into practice on January 1, 1981. The document stipulates three academic degrees: bachelor, master and doctoral. College graduates with outstanding academic achievements, well grounded in basic theory and practice, and able to do research or technical work in their chosen fields, will be granted the bachelor degree. Postgraduates from institutions of higher learning, including research institutes, and people with an academic record corresponding to that of postgraduates, well grounded in basic theories and having systematic knowledge in their specialized chosen fields, being able to do independent research or special technical work, and having successfully passed the required examinations of all the relevant courses and defended the theses, will be granted the master degree. A master-degree holder, being proficient in theory and practice of the subject matter of his choice, being able to do research work on his own, and having been credited with creative academic achievement, could be granted the doctoral degree, if he has, in the meantime, succeeded in passing the examinations of relevant courses and in defending his dissertation. Only a State Council authorized college or university and institute of research can award bachelor, master and doctoral degrees. Institutions of higher learning and research institutes must seek the State Council's approval before they can grant academic degrees in specified academic disciplines. The establishment of academic degrees will have a positive effect on the training of specialized personnel and on the development of education in general and science in particular.

Academic Exchange   In recent years institutions of higher learning have strengthened the work of international academic exchange. By the end of 1979, 108 Chinese institutions of higher learning had established contacts with 130 sister institutions in more than 30 countries in the exchange of scholarly papers, books and library materials, professors and students for study, research, giving or attending lectures, or short-term observation. In 1979, Chinese institutions of higher learning sent more than 2,700 of their undergraduate and postgraduate students to 41 countries. Early in 1979 there were 1,258 foreign students (including students for advanced study) in China from 50 countries. In the same year, 370 experts were invited from more than 30 countries to teach foreign languages in China, in addition to more than 250 foreign scholars who had been invited to China to give lectures on science and technology.

Meanwhile, institutions of higher learning in China have sent an increasing number of professors to give lectures, observe, or take part in international academic conferences. In 1979, Chinese institutions of higher learning sent 119 delegations abroad for the purpose of observation and study and 106 delegations to take part in international academic conferences.

Teaching and Administrative Structure The president is the chief administrator of a college or university appointed by the State Council. All important decisions must be submitted to the school's Party committee for approval before they are carried out by the president. Inside a university or college there are various administrative organs in charge of personnel, teaching, research, productive labour and general affairs.

Inside each institution of higher learning there are various academic departments, the chairmen and vice-chairmen of which are appointed by the president. In each department there are teaching and research committees, reference rooms, and laboratories. The head and deputy head of each teaching and research committee, who can be either elected or appointed, are responsible for the implementation of the teaching plan, selecting and compiling teaching materials, working out syllabi, and supervising research work.

On the college or university level, there are also various academic committees composed of experienced experts and professors. In addition to the making of recommendations regarding the institution's educational development, scientific research and training of postgraduates, they evaluate the results of research, assess

Ninety-seven Key Colleges and Universities

Name	Type	Location
* Chinese People's University	comprehensive	Beijing
* Beijing University	"	"
* Qinghua University	science and engineering	"
Jiaotong University in North China	"	"
Beijing Aeronautical Engineering Institute	"	"
Beijing Engineering Institute	"	"
Beijing Iron and Steel Engineering Institute	"	"
Beijing Posts and Telecommunications Institute	"	"
Beijing Chemical Engineering Institute	"	"
Beijing Agro-technical Institute	"	"
Beijing Agricultural University	agriculture	"

Name	Type	Location
Beijing Forestry Institute	forestry	"
China Capital Medical University	medicine	"
Beijing Medical College	"	"
Beijing Traditional Chinese Medicine Institute	"	"
* Beijing Normal University	teacher training	"
Beijing Foreign Languages Institute	language	"
Central Institute for Nationalities	nationality	"
Beijing Institute of Foreign Trade	business administration and economics	"
College of International Relations	political science and law	"
Beijing Physical Culture Institute	physical culture	"
Central Conservatory of Music	art	"
* Fudan University	comprehensive	Shanghai

Name	Type	Location
* Tongji University	science and engineering	"
Jiaotong University in Shanghai	"	"
* East China Chemical Engineering Institute	"	"
Shanghai Textile Engineering Institute	"	"
Shanghai No. 1 Medical College	medicine	"
* East China Normal University	teacher training	"
* Shanghai Foreign Languages Institute	language	"
* Nankai University	comprehensive	Tianjin
* Tianjin University	science and engineering	"
* Nanjing University	comprehensive	Nanjing
* Nanjing Engineering Institute	science and engineering	"
East China Water Conservancy Institute	"	"

Name	Type	Location
Nanjing Meteorological Institute	"	"
Nanjing Aeronautical Engineering Institute	"	"
East China Engineering Institute	"	"
Zhenjiang Agromachinery Institute	"	Zhenjiang, Jiangsu
Nanjing Agricultural College	agriculture	Nanjing
* Zhejiang University	science and engineering	Hangzhou, Zhejiang
Chinese University of Science and Technology	"	Hefei, Anhui
Hefei Engineering University	"	"
* Xiamen University	comprehensive	Xiamen, Fujian
* Zhongshan (Sun Yat-sen) University	"	Guangzhou, Guangdong
* South China Engineering Institute	science and engineering	"
South China Agricultural College	agriculture	"

Name	Type	Location
Zhongshan (Sun Yat-sen) Medical College	medicine	"
* Shandong University	comprehensive	Jinan, Shandong
* Shandong Oceanography College	science and engineering	Qingdao, Shandong
East China Petroleum Institute	"	Dongying, Shandong
North China Institute of Electric Power	"	Baoding, Hebei
Inner Mongolia University	comprehensive	Huhhot, Inner Mongolia
* Dalian Engineering Institute	science and engineering	Dalian, Liaoning
Dalian Mercantile Marine Institute	"	Dalian, Liaoning
Northeast China Engineering Institute	"	Shenyang, Liaoning
Shenyang Agricultural College	agriculture	"
Fuxin Mining Institute	science and engineering	Fuxin, Liaoning
* Jilin University	comprehensive	Changchun, Jilin

Name	Type	Location
Jilin Poly-technic Institute	science and engineering	"
Changchun Geological Institute	"	"
Harbin Engineering Institute	"	Harbin, Heilongjiang
Harbin Ship Engineering Institute	"	"
Northeast China Heavy Machinery Institute	"	Fulaerji, Heilongjiang
Daqing Petroleum Institute	"	Anda, Heilongjiang
Jiangxi Agricultural University	agriculture	Nanchang, Jiangxi
Xiangtan University	comprehensive	Xiangtan, Hunan
Hunan University	science and engineering	Changsha, Hunan
University of National Defence Science and Technology	"	"
Central-South China Institute of Mining and Metallurgy	"	"

Name	Type	Location
* Wuhan University	comprehensive	Wuhan, Hubei
* Central China Engineering Institute	science and engineering	"
Wuhan Institute of Water Conservancy and Electric Power	"	"
Wuhan Institute of Cartography	"	"
Wuhan Geological Institute	"	"
Wuhan Building Materials Engineering Institute	"	"
Central China Agricultural College	agriculture	"
Xinjiang University	comprehensive	Urumqi, Xinjiang
* Lanzhou University	"	Lanzhou, Gansu
Northwest China Agricultural College	agriculture	Wugong, Shaanxi
Northwest China Light Industry Institute	science and engineering	Xianyang, Shaanxi

Name	Type	Location
Northwest China Telecommunications Engineering Institute	"	Xian, Shaanxi
Northwest China Engineering Institute	"	"
* Jiaotong University in Xian	"	"
Northwest China University	comprehensive	"
Yunnan University	"	Kunming, Yunnan
* Sichuan University	"	Chengdu, Sichuan
* Chongqing University	science and engineering	Chongqing, Sichuan
* Chengdu University of Science and Technology	"	Chengdu, Sichuan
Jiaotong University in Southwest China	"	Emei, Sichuan
Chengdu Telecommunications Engineering Institute	"	Chengdu, Sichuan
China Mining Institute	"	Hechuan, Sichuan

Name	Type	Location
Chongqing Architectural Engineering Institute	"	Chongqing, Sichuan
Southwest China Agricultural College	agriculture	"
Sichuan Medical College	medicine	Chengdu, Sichuan
Southwest China Institute of Political Science and Law	political science and law	Chongqing, Sichuan
Shanxi Agricultural University	agriculture	Taigu, Shanxi

* Key institutions of higher learning under the direct jurisdiction of the Ministry of Education.

postgraduates' theses or dissertations, determine the promotion of professors and associate professors, sponsor academic forums, and select participants in academic exchanges at home or abroad.

EIGHTEEN INSTITUTIONS OF HIGHER LEARNING

Beijing University

Beijing University, located in a scenic spot in the northwestern suburb of Beijing, is a well-known comprehensive university covering humanities and sciences. Formerly the Metropolitan College founded in 1898, it acquired its present name in 1912. During the period of the War of Resistance Against Japan, it was moved to Kunming, where, jointly with Qinghua and Nankai universities, it became known as the Southwest Union University. In 1946 it returned to Beijing. After Liberation, changes were introduced to the institutions of higher learning, and Beijing University was moved in 1952 to the campus of the former Yanjing University. The role of the university is to train specialized personnel needed by the state for scientific research and teaching in such branches of learning as philosophy, social sciences, natural sciences, languages, and literature.

Beijing University has a glorious revolutionary history. Li Dazhao, one of the founders of the Chinese Communist Party, had organized the country's first Marxist study society in the university. In 1918 Mao Zedong worked in the library of Beijing University. Lu Xun also worked there at one time. Beijing University was the place where the epoch-making May 4 Movement of 1919 began.

At present, Beijing University has such academic departments as philosophy, history, economics, law, international politics, library science, Chinese language and literature, Russian language and literature, Oriental languages and literature, Western languages and literature, mathematics, mechanics, radio and electronics, geophysics, technical physics, computer science and technology, physics, chemistry, biology, geology, geography and psychology.

The university has also research institutes on such subject matters as mathematics, solid state physics, theoretical physics, large ion physics, physical chemistry, molecule biology, computer science, remote sensing technology, foreign philosophy, and South Asian studies.

The university has attained a high level of excellence in many subjects of study. Take its department of theoretical physics as an example. It is well known throughout the country, and many bright, ambitious young people wish to study there. The departments of mathematics and mechanics are also solid; young mathematicians Yang Le and Zhang Guanghou both are their products. Outstanding achievement has been made by the biology department that took part in the research of synthetic crystalline bovine insulin.

The university has a library holding of 3,200,000 volumes, in addition to a large rare book collection. Being the largest among China's institutions of higher learning, the Beijing University library has 31 reading rooms with a total of 2,400 seats.

During the academic year 1979-80, the university had a student body of more than 8,000, of whom more than 7,000 were undergraduates, 600 were postgraduates and 359 were students for advanced study. There were more than 2,700 teachers and scientific research personnel; professors and associate professors numbered 405 and lecturers 1,747.

Having made contacts with many foreign universities, Beijing University not only sent its own students to study abroad but also invited foreign experts and professors to give lectures in China. In 1980, it had more than 150 foreign students from more than 30 countries.

Among the Beijing University publications are *Beijing University Journal* (*Natural Sciences*) and *Beijing*

University Journal (Philosophy and Social Sciences). Both are circulated nationwide.

Qinghua University

Founded in 1911, Qinghua University is located in the western suburb of Beijing. At one time, it was a comprehensive university where famous scholars like Wen Yiduo, Zhu Ziqing and Liang Sicheng once taught. When adjustments were made after Liberation, it was changed into a poly-technic university. Having a strong faculty and a good foundation in both teaching and research, it is famous throughout the country. Its length of study is five years.

The university has the following academic departments: construction engineering, civil and environmental engineering, water conservancy engineering, mechanical engineering, precision instruments, thermal energy engineering, electric power engineering, industrial automation, business administration, applied physics, chemistry and chemical engineering, radio and electronics, computer engineering and science, and engineering mechanics. In addition to a research institute on teaching methodology, there are also research institutes for various academic disciplines.

In 1980, the university had more than 7,000 postgraduate and undergraduate students. It had also about 100 foreign students (including those for advanced study) from more than 22 countries. The number of full-time teachers totalled 3,600, of whom more than 600 were professors and associate professors, and more than 1,600 were lecturers.

The university has 87 laboratories for teaching and research. Its library holding totals 2,000,000 volumes. Every academic department has its own reference room.

The university publishes the *Qinghua University Journal* which has a nationwide circulation.

Chinese University of Science and Technology

Founded by the Chinese Academy of Sciences in 1958 at Beijing, the Chinese University of Science and Technology is a university that integrates science with technology. In 1970, it was moved to Hefei, Anhui Province. Guo Moruo, former president of the Chinese Academy of Sciences, famous scientist and man of letters, was at one time its president. Eminent scientists like Zhu Kezhen, Wu Youxun, Yan Jici, Hua Luogeng and Qian Xuesen have taught or been placed in leading positions in the university. Its present president is the famous physicist Yan Jici.

The main task of the university is to train scientific research personnel; it also trains engineers and technicians and personnel for scientific management. The length of study is five years.

The university has such academic departments as mathematics, physics, modern chemistry, modern physics, modern mechanics, radio and electronics, earth and space science, biology and precision machinery.

Since 1977 the university has opened three classes to 117 gifted youngsters of less than 15 years in age, including 16 girls. They were selected from recommendees from various provinces and municipalities as high achievers who excelled in most subjects of study, especially mathematics and science.

At present the university has a teaching staff of 1,426, of whom 87 are professors and associate professors, and 734 are lecturers. In 1980 the university had a student body of more than 3,000, including 142 postgraduates.

Each academic department of the university trains students jointly with a research institute of the same academic interest. The length of undergraduate study is five years, the first three and half for the enrollment in basic courses, and the last one and half for the studying of more advanced specialized courses. Some students have to write theses under the guidance of relevant research institutes before their graduation.

In Beijing the university has a postgraduate institute with a student body of 1,200. Its length of study is three years. Half of the students' time in the three years is devoted to the study of basic courses or interdepartment courses which all students must take, and the other half of the time is taken up by research work under the guidance of a tutor from a research institute.

The university has 40 laboratories, in addition to one equipped with advanced electronic synchrotron radiation accelerator currently under construction. Its library has a book holding of more than 500,000 volumes, in addition to 200,000 journals and magazines.

With the approval of the State Council, the university, to encourage students to scale the heights of science, has established a "Guo Moruo Scholarship" to reward students who love socialism and are especially outstanding in academic work.

The university publishes the *Chinese University of Science and Technology Journal*.

Nankai University

Founded in 1919 and located in Tianjin, Nankai University is a comprehensive university covering humanities and sciences.

During the early years, the university offered courses on liberal arts, science and business administration. Then it added courses on mining, chemical engineering and electrical engineering. In 1927, it established the Nankai Economic Research Institute, and in 1932, the Applied Chemical Research Institute. After Liberation it became a comprehensive university covering humanities and sciences. Even today the university is still well known throughout the country for the study of economics and chemistry.

Nankai University is also known for its revolutionary and democratic tradition. Zhou Enlai was one of its first graduates majoring in liberal arts. While a student here during the May 4 Movement, he led the people of Tianjin in the patriotic movement to fight against imperialism and feudalism and founded the Awakening Society.

Today the university has such academic departments as Chinese language and literature, foreign languages and literature, history, political economy, philosophy, mathematics, physics, chemistry and biology.

The university has four research institutes and 12 research departments. They are: institute of organic chemistry, institute of molecular biology, institute of economics and institute of mathematics, department of modern optics, department of solid energy spectrum, department of theoretical physics, department of structural theory on organic chemistry, department of entomology, department of genetics, department of American history, department of Japanese history, department of Ming and Qing histories and department of Zhou Enlai.

The university's library holding has increased from 100,000 volumes in the early post-Liberation period to 1,200,000 volumes.

In 1979, the university had 3,684 undergraduate students and more than 200 postgraduates. It had 1,465 persons engaged in teaching and scientific research, among whom 116 were professors and associate professors, and 654 were lecturers.

The university publishes *Nankai University Journal* that has a nationwide circulation. The publication has three editions: natural science, philosophy and social sciences.

Fudan University

Fudan University is a comprehensive university covering humanities and sciences. It has had a history of 74 years since its founding in Shanghai in 1905.

The university has such academic departments as Chinese language and literature, foreign languages and literature, journalism, history, international politics, political economy, philosophy, mathematics, physics, chemistry, optics, biology, computer science, management science and international economy.

In addition to more than 30 research departments, the university has the following research institutes: institute of mathematics, institute of genetics, institute of modern physics, institute of electric light and institute of international economy.

The university owns an electronic instrument factory and a petro-chemical plant.

In 1979, the university had more than 2,000 teachers, of whom 200 were professors and associate professors and 800 were lecturers. It has a student body of more than 4,000, including more than 200 postgraduates and 25 foreign students.

The library of the university has a book holding of 1,650,000 volumes.

The university publishes *Fudan University Journal* (*Social Science*) and *Fudan University Journal* (*Natural Science*).

Zhongshan (Sun Yat-sen) University

Formerly known as Guangdong University, and founded by the Chinese democratic revolutionary Dr. Sun Yat-sen in 1924, it acquired its present name in 1926 as a tribute to its founder.

Located in Guangzhou, it is a comprehensive university covering humanities and sciences. It has such academic departments as Chinese language and literature, history, philosophy, economics, law, foreign languages and literature, mathematics, mechanics, physics, chemistry, biology, geography, geology, computer science, radio and electronics, and meteorology.

The university has six research institutes of history of Southeast Asia, mathematics, computer science, high polymer, environmental science and entomology. In addition, it has 18 research departments of Dr. Sun Yat-sen, Lu Xun, old scripts, history of modern and contemporary international relations, history of modern and contemporary Chinese philosophy, theory of elementary particle, physics of gravitational force, laser optics and spectroscophy, submillimeter wave, solid physics, hydrodynamics, catalytic theory, natural organic chemistry, delta and seacoast, ichthyology, botany, genetics and parasitology.

In 1979, the university had 1,071 teachers, of whom 117 were professors and associate professors, and 703 were lecturers. It had a student body of 4,129, of whom 284 were postgraduates and 34 were foreign students.

The university has 120 laboratories. Its library has a book holding of 2,400,000 volumes. It publishes *Zhongshan University Journal* (*Natural Science*) and *Zhongshan University Journal* (*Philosophy and Social Science*).

Wuhan University

Founded in 1913, Wuhan University is a comprehensive university covering humanities and sciences. It is located at the foothill of the Luojia Mountains, next to the scenic Donghu Lake in the suburb of Wuhan.

The university has the following academic departments: Chinese language and literature, history, philosophy, political economy, law, library science, foreign languages and literature, French language and literature, computer science, physics, space physics, chemistry, biology and virology.

The university has nine research institutes of mathematics, computer science, virology, solid physics, radio wave transmission and space physics, bionics chemistry, natural dialecticism, problems in foreign countries, and medieval Chinese literature, history and philosophy. In addition, it has 44 research departments studying such subjects as electrochemistry, modern analytic chemistry, genetics, philosophy of Mao Zedong, theory on literature and art, history of Wei, Jin and the Southern and Northern Dynasties, history of the United States, economy of North America and foreign literature.

In 1980, the university had 1,625 teachers, of whom 235 were professors and associate professors, and 731 were lecturers. In addition, there were also invited foreign teachers. Then the university had a student body of 4,385, of whom there were 294 postgraduates. In recent years more than 50 undergraduates and postgraduates have been sent to study abroad.

The university has more than 90 laboratories, 1 computer centre and 1 analytic and test centre. It has also 6 experimental and service factories.

The university library has a total holding of 1,500,000 volumes, of which 460,000 are foreign books and magazines. There is also a centre of foreign teaching materials with a collection amounting to more than 5,800 in variety.

The university publishes for nationwide circulation the *Wuhan University Journal* (*Natural Science*) and the *Wuhan University Journal* (*Philosophy and Social Science*).

Beijing Normal University

One of the oldest institutions of higher learning for teacher training in the country, it was formerly known as the Normal School of the Metropolitan College, which was founded in 1902. After the Revolution of 1911, it was renamed Beijing Higher Normal School. In 1908, the Beijing Women's Normal School was established. In 1918, it was renamed Beijing Women's Higher Normal School. In 1931, the Beijing Higher Normal School and the Women's Higher Normal School merged to become the Beijing Normal University, located in Beijing.

The main task of the university is to train teachers for middle schools as well as research personnel. The university has 14 academic departments offering 19 fields of study. The academic departments are those of education, philosophy, political economy, history, Chinese language and literature, foreign languages and literature, mathematics, physics, astronomy, chemistry, biology, geography, physical culture, radio and electronics.

The university has 6 research institutes specializing in science of education, foreign education, history, Soviet literature, low energy nuclear physics, and methodology of modern education.

In 1979 the university had a faculty of 1,285 for teaching and research; among them were 141 professors and associate professors, and 657 were lecturers. In addition, there were invited foreign teachers from Japan, the United States, Australia and Canada. The university had then 3,000 undergraduates and more than 300 postgraduates and students for advanced study.

The university publishes the *Beijing Normal University Journal* which has two editions, one for social science and one for natural science. It also publishes *Teaching Bulletin* and *Foreign Languages Teaching in Primary and Middle Schools*. With a book holding of more than 2,000,000 volumes, its library serves as the centre of learning in education throughout the country. The university has three affiliated middle schools, and all three are key schools in Beijing.

Beijing Medical College

Founded in Beijing in 1912, it was formerly the Medical College of Beijing University. Since Liberation, it has been independent.

The college has five academic departments, namely, medical science, basic medical science, sanitation and public hygiene, stomatology and pharmacology (the last containing two disciplines: chemical pharmaceutics and chemistry). It has four hospitals with a total of more than 2,000 beds. Within its jurisdiction are a medical school and a pharmaceutical factory.

The college has research institutes specializing in medical science education, basic medical science, environmental medical science, pharmaceuticals, clinical medical science, clinical pharmacology, stomatology, urological surgery, prevention and treatment of cancer, blood disease, sports medical science, and psychiatrical hygiene. There are also research departments specializing in plastic surgery, occupational disease, disease of the digestive system, mental illness, surgical cancer and liver disease.

During the first half of 1980 the college had 1,600 teachers and physicians, of whom 158 were professors and associate professors, and approximately 1,000 were lecturers and chief physicians. It had 2,000 undergraduates, 160 postgraduates, 600 students for advanced study, and about 100 foreign students.

The college library has a book holding of 360,000 volumes.

The college publishes the *Beijing Medical College Journal* and *Education of Medical Science*, both with nationwide circulation.

Jiaotong University

The Jiaotong University in Xian is a poly-technic institution.

Founded in Shanghai in 1896, Jiaotong University has had a history of 85 years. To meet the requirement for rational distribution of industry and socialist construction, the university in 1956 moved most of its students and equipment from Shanghai to Xian. In 1959, the Xian branch of the university was named Jiaotong University in Xian.

The university has the following academic departments: mechanical engineering, power engineering, electrical engineering, electronics engineering, information and control engineering, mathematics, engineering mechanics, and basic courses.

The university has research institutes specializing in metals and metal strength, systems engineering, mechanical engineering and thermal physics. It has three research departments studying electrical insulation, electrical engineering and computer science and technology.

Attached to the university are a teaching equipment factory, a practice factory, a radio factory and a printing shop.

In 1980, the university had a teaching staff of nearly 1,500, of whom 218 were professors and associate professors, and 912 were lecturers. It had a student body of 6,520, including 184 postgraduates. In addition, it has had an extensive academic exchange programme with foreign countries, and many famous foreign scholars have visited it. Today there are 16 foreign experts teaching in the university.

The university library has a book holding of 1,000,000 volumes. It can seat 1,500 readers simultaneously.

The university publishes the *Journal of Jiaotong University in Xian* which has a nationwide circulation.

Shandong University

Founded in Qingdao in 1926, Shandong University is a comprehensive university covering humanities and sciences. In 1958 it was moved to Jinan, provincial capital of Shandong.

Shandong University has definitely exercised much influence in academic fields, especially in the field of literature, history and biology. Famous scholars like the poet Wen Yiduo, the writers Wang Tongzhao and Lao She, and the biologist Tong Dizhou once taught in the university. The university publishes *Literature, History and Philosophy,* a theoretical journal that exercises great influence across the country.

In the university there are such academic departments as Chinese language and literature, history, philosophy, scientific socialism, political economy, foreign languages and literature, mathematics, physics, chemistry, radio and electronics, optics and biology.

The university has five research institutes specializing in crystal material, microorganism, infrared remote sensing, quantum chemistry and elementary particle.

In 1980, the university had a teaching staff of 1,150, of whom 116 were professors and associate professors and 432 were lecturers. There was a student body of 4,500, of whom 250 were postgraduates.

The university has a library with a book holding of 1,500,000 volumes. It publishes *Shandong University Journal* (*Natural Science*) and *Literature, History and Philosophy*.

Central Institute for Nationalities

Founded in Beijing in 1951, the Central Institute for Nationalities is the highest educational institution for minority nationalities in the country.

The main task of the institute is to train political cadres and specialized technical personnel for the minority nationalities. Over the years, more than 14,000 graduates have left the institute, including some 200 postgraduates and foreign students. In 1979, the institute had a student body of more than 2,200 from over 50 nationalities, and a teaching staff of 673, of whom 62 were professors and associate professors and 329 were lecturers. It had 25 postgraduates.

The institute has the following academic departments: political science, Han language, history, languages of minority nationalities, mathematics, physics and art. It has an in-service centre for the training of minority nationality cadres, and the length of training is one year. It also runs a preparatory school for minority students who wish to participate in college entrance examination, and the length of schooling is one to two years.

The college has four research institutes specializing in the theory and policies on minority nationalities, their social history, spoken and written languages, art and literature and economy. Its library has a book holding of more than 700,000 volumes.

Teachers and students in the institute are free to use their own languages, wear their national costumes and celebrate their traditional festivals. The institute operates a Muslim canteen for teachers and students of ten minority nationalities. At all times, consideration and respect are accorded to the custom and tradition of all minority nationalities.

The institute publishes the *Journal of the Central Institute for Nationalities.*

Beijing Foreign Languages Institute

The main task of the institute is to train foreign language translators and teachers. The institute is located in Beijing.

Formerly the Russian Language Section of the Chinese People's Anti-Japanese Military and Political Academy of Yanan, it was renamed Yanan Foreign Languages School in 1944, when for the first time it offered courses in English. In 1948, two schools emerged, the Foreign Affairs School with concentration in English and the Beijing Russian Language School. Later, the former was renamed Beijing Foreign Languages Institute, and the latter, Beijing Russian Language Institute. When the two merged in 1959, the result was known as Beijing Foreign Languages Institute.

The institute has the following academic departments: English, Russian, French, Spanish, Asian and African languages, and Eastern European languages.

The institute operates a postgraduate school, a school for advanced students and a foreign languages school for primary and secondary school students. In 1980, at the request of the United Nations, the institute began to operate a programme for the training of simultaneous interpreters and translators. It also teaches English via radio and television.

In 1979, the institute had a teaching staff of 590, of whom 74 were professors and associate professors, and 293 were lecturers, in addition to 49 foreign experts and teachers who came from 19 foreign countries and regions.

The institute has 983 undergraduates, 82 postgraduate students, including students in translators' classes and classes for advanced studies, and 16 foreign students. Its foreign languages school has 895 students.

The institute subscribes 278 newspapers and magazines. Its library has a book collection of 410,000 volumes, of which 230,000 volumes are in foreign languages. It publishes *Foreign Languages Teaching and Study*, *English Learning* and *German Learning*. Among its other publications are six Chinese-English dictionaries and three classified Chinese-English and Chinese-German glossaries. At the moment it is working on 17 dictionaries in Spanish-Chinese, Chinese-German, etc.

Beijing Languages Institute

The main task of the Beijing Languages Institute is to teach Chinese to foreign students in China. It also teaches Chinese students foreign languages and other subjects.

Formerly a section of Qinghua University for teaching foreign students Chinese, it was incorporated into Beijing University in 1952 and into Beijing Foreign Languages Institute in 1961. Due to the ever-increasing number of foreign students in China, it became an independent institute in 1962. Known at first as Beijing Advanced Preparatory School for Foreign Students, it was renamed Beijing Languages Institute in 1964.

The institute has the first and second departments for foreign students, the department of foreign languages, the preparatory department for students going to study abroad, and the research and editorial department.

The first department for foreign students offered basic courses of the Chinese language. After finishing their study in this department, foreign students are transferred to other institutions of higher learning to study courses of their chosen fields.

The second department for foreign students trains translators and teachers of modern Chinese and accepts those foreign students who have already attained a certain level of proficiency in Chinese.

Foreign students majoring in science, engineering, agriculture and medical science will study Chinese for one year in the institute. The length of study is two years for foreign students majoring in literature, history, philosophy and traditional Chinese medicine and four years for foreign students majoring in modern Chinese.

The main task of the department for foreign languages is to train translators and teachers of foreign languages. Currently six foreign languages are taught: English, French, Japanese, German, Arabic and Spanish. The length of study is four years, and students are Chinese.

The preparatory department for students going to study abroad prepares scientists, technicians, and students in foreign languages before they go abroad. Four courses are taught at the moment: English, French, German and

Japanese. The study lasts anywhere from one half to one year.

The research and editorial department studies the best way to teach foreigners Chinese, and it writes and compiles books for foreigners who study Chinese. It publishes the journal *Teaching and Study of Chinese*, circulated both at home and abroad.

During the academic year 1979-80, the institute had 350 Chinese students and 600 scientists and technicians due to leave China for studying abroad, 447 foreign students from 53 countries, and a teaching staff of 509, of whom 5 were professors and associate professors, 266 were lecturers, and 18 were foreign experts from 7 countries.

Beijing Iron and Steel Engineering Institute

Beijing Iron and Steel Engineering Institute is a metallurgical engineering and technological school. It came about in 1952 when the metallurgical and mining departments of Beiyang University, Tangshan Engineering Institute and Northwest Engineering Institute merged. At the beginning, the institute had only 40 or more teachers and approximately 400 students. Now it is one of the three key metallurgical and mining institutes in the country.

The institute has six academic departments specializing in mining, metallurgy, metals, machinery, automation and physical chemistry, in addition to a department that teaches basic courses. In 1980, it had a teaching staff of 1,149, of whom 160 were professors and associate professors and 846 were lecturers. There were 3,614 students, of whom 268 were postgraduates.

The institute runs a practice factory, laboratories, and a centre of material-structure analysis, a computer centre, a centre of chemical examination and analysis, a research and manufacturing centre of instruments and meters, and a printing house.

The institute library has a book holding of more than 700,000 volumes.

University for Overseas Chinese

Founded in 1960 and located at Quanzhou of Fujian Province, the University for Overseas Chinese is a comprehensive university for young overseas Chinese. In 1965 it began to grow in size. It then had five academic departments: liberal arts, science, engineering, agriculture and medical science. It had 2,300 students, over 90 per cent of whom were young Chinese from abroad. The university was closed during the "cultural revolution" but reopened in the spring of 1978. Now under the direct jurisdiction of the Ministry of Education, it stresses training in engineering and, to a lesser extent, science. Liao Chengzhi, Vice-Chairman of the Standing Committee of the National People's Congress and Director of the Office of Overseas Chinese Affairs, is now its president. Most of its students are young overseas Chinese, young people from Hongkong, Macao and Taiwan, and children of returned overseas Chinese and their relatives. After graduation, students coming from abroad and those from Hongkong and Macao can either remain in the country where they will be given work assignments by the state or return to countries or regions of their origin to seek employment on their own.

The university has six academic departments: physics, mathematics, chemistry, civil engineering, chemical engineering and mechanical engineering. The length of study is four years. At present there are some 800 students, 391 faculty members and assistants, in addition to a total of 285 staff members. The planned enrollment in the near future will be about 2,000.

The medium of instruction in the university is the standard spoken Chinese. Whenever necessary, instruction will be given wholly or partly in English. Remedy courses are available to those who are inadequately prepared in Chinese.

The university is supplied with an increasing amount of equipment for its laboratories and books and reference materials for its library. It has received generous contributions from patriotic Chinese abroad and compatriots in Hongkong and Macao. The library has a book holding of more than 400,000 volumes.

Central Academy of Fine Arts

Founded in 1918, the National Beiping Art School merged with the fine arts department of the North China Union University (located formerly in the liberated areas) to become the Central Academy of Fine Arts after Liberation.

There are five departments in the academy: Chinese traditional painting, oil painting, graphic art, sculpture, and history of fine arts. A middle school of fine arts and a creative sculpture studio are attached to the academy.

The academy publishes two journals: *Study of Fine Arts* and *Fine Arts of the World.*

The academy has 91 undergraduates, 65 postgraduates, and 9 foreign students. It has a teaching faculty of 140, of whom 44 are professors and associate professors and 13 are lecturers.

Xiamen University

Founded in 1921 with funds provided by a patriotic overseas Chinese named Tan Kah-kee, Xiamen University became a national institution of higher learning in 1937. It became a comprehensive university covering humanities and sciences after Liberation. It has in it an overseas correspondence school which, before the "cultural revolution", trained by correspondence more than 10,000 overseas Chinese students in more than 30 countries and regions of the world.

The university has ten academic departments: Chinese language and literature, foreign languages and literature, history, economics, philosophy, mathematics, physics, chemistry, biology and oceanography. The university has also 16 research institutes, specializing in physics and chemistry, oceanography, biology, etc.

In 1980, the university had 4,513 undergraduates and 156 postgraduates. It had 900 teachers and 190 full-time scientific research personnel, of whom 92 were professors, associate professors, research and associate research fellows, and 726 were lecturers and assistant researchers.

The university library has a book holding of more than 1,000,000 volumes.

Located on the seaside and picturesque, it has a natural beach for swimming and a large seawater swimming pool.

Region	Number of Students / Year	Primary Schools	Junior Middle Schools	Senior Middle Schools	Institutions of Higher Learning
Tibet Autonomous Region	1951	0	0	0	0
	1978	236,000	15,200	1,900	2,081
Ningxia Hui Autonomous Region	1951	56,000	1,000	100	0
	1978	630,000	184,800	55,800	2,890
Xinjiang Uygur Autonomous Region	1951	307,000	2,900	300	476
	1978	2,029,000	677,000	129,000	10,275
Guangxi Zhuang Autonomous Region	1951	1,580,000	55,300	5,700	2,235
	1978	5,117,000	1,648,900	562,300	21,079
Inner Mongolia Autonomous Region	1951	590,000	7,900	900	0
	1978	1,374,000	669,500	120,600	9,895

The university publishes the *Journal of Xiamen University* (*Philosophy and Social Science*) and *Journal of Xiamen University* (*Natural Science*).

(6) EDUCATION FOR MINORITY NATIONALITIES

Brief Introduction   To develop minority nationality areas politically, economically and culturally and train qualified minority personnel in various spheres of endeavour, various types of schools have been established in areas inhabited by minority nationalities since the founding of the People's Republic.   There has been a speedy development of education in these areas.

Meanwhile, the proportion of students of minority nationalities in the total number of students in the country has increased considerably.

	Institutions of Higher Learning	Secondary Technical Schools	Secondary Normal Schools	Ordinary Middle Schools	Primary Schools
1978	4.21%	5.4%	8.4%	3.8%	5.3%
1951	1.36%	0.4%	2.1%	2.6%	2.2%

In 1978, the proportion of minority nationality teachers in the total number of teachers in the country was 2.85 per cent in institutions of higher learning, 3.9 per cent in secondary technical schools, 6.5 per cent in secondary normal schools, 3.5 per cent in ordinary middle schools and 5.9 per cent in primary schools.

Middle and Primary Schools for Minority Nationalities In terms of purpose, curriculum and teaching plans, middle and primary schools for minority nationalities are basically the same as those of the Han schools. There are, however, differences. First, in schools of the former type, a minority student must study his own language as well as Chinese, and he is required to be proficient in both. More and more time will be devoted to the study of Chinese as he grows older and older. Second, the length of study in schools of the former type is usually one year longer than that of the latter type owing to the fact that students have to study two languages. Third, instruction in schools for minority nationalities is given in the languages of these nationalities. Wherever conditions permit, nationality schools are established. In areas that have a small number of minority nationality students, nationality classes will be held in ordinary schools. In multi-nationality classes, instruction will be given in the language used by most of the people in the area. Fourth, characteristics of the areas where minority nationalities reside must be taken into consideration when determining how schools should be run. For instance, in pastoral, sparsely populated, remote or mountainous areas, schools, generally run by the state, provide both room and board.

Minority Students in Colleges and Universities Colleges and universities show special care for students of minority nationalities in a variety of ways. Over the last few years, they lowered the academic standard for admission as far as minority students are concerned. In 1979, the colleges and key universities also accepted these students even though they scored far fewer points than the normally successful candidates. The Ministry of

Education has stipulated that for admission to institutions of higher learning in various national autonomous regions where instruction is given in minority languages, students do not have to take the unified, nationwide entrance examinations. Educational authorities in these regions can make their own questions and determine their own standards for accepting or rejecting candidates who, in fact, can use their own languages to answer questions in the entrance examination. To enable students of minority nationalities to study in better schools, a number of key colleges and universities have opened special classes for them to study Chinese and other basic subjects. These students are given consideration and care by teachers and school leaders concerning their life and study. Generally speaking, they will be assigned to work in their home communities after graduation.

Institutes for Minority Nationalities To train political cadres, teachers, and technical and specialized personnel to work among minority nationalities, the state has established ten institutes for minority nationalities in Beijing, southwest China, central China and northwest China. These institutes are different not only from ordinary comprehensive universities but also from other cadres' schools for minority nationalities. Yet they possess the characteristics of both. Apart from courses on liberal arts and sciences, they also train political cadres and offer prerequisite courses for college aspirants. Some of them have such academic departments as languages and arts of minority nationalities. Others located in the border areas where educational and cultural level is extremely low may offer primary and middle school courses as well as those of college. Still others may open special classes in order to raise the students' proficiency in the

Chinese language. The length of study varies. It takes one year to train political cadres, one to two years to prepare students for college, and four years for a baccalaureate. Some institutes offer research programmes for the study of the history, policy, and languages of minority nationalities.

Students in any of these institutes are at liberty to speak their own minority languages, wear their own national costumes and celebrate their own festival. Their custom and way of life are always respected.

Since the founding of the People's Republic, 94,000 students of 56 nationalities (including a small number of the Han nationality) have been trained by the various institutes for minority nationalities. Over 10 per cent of the cadres of various nationalities are graduates of the country's institutions of higher learning for minority nationalities. In 1978, these institutions of higher learning had a student body of 9,100 and a full-time faculty of 2,121, of whom 610 or 28 per cent, were members of minority nationalities.

Training of Teachers To increase the number of teachers for minority nationalities and to raise their professional level, the state not only has established new normal colleges and schools in regions where minority nationalities reside, but also each year sends a fixed number of college graduates from other parts of the country to these regions to work as teachers. Over the last few years, the state has organized many teachers to work in the remote areas where minority nationalities reside, usually on a short-term basis. Many institutions of higher learning in the hinterland also train teachers for minority nationalities.

(7) SPECIAL EDUCATION

Brief Introduction   Special education means education for the blind, deaf-mutes, people low in intelligence
or having physical defects.   Since the founding of the
People's Republic, the government has paid special attention to special education and integrated it into the general
educational system.   Today there are three kinds of special education schools, namely schools for the blind,
schools for the deaf-mutes and schools for the blind and
the deaf-mutes.   Students with physical defects or low
in intelligence are mostly enrolled in ordinary schools,
which give them special care and attention.   Recently a
special school for the mentally retarded has been established in Shanghai on an experimental basis.   Shanghai
has opened "supplementary classes in its ordinary primary schools for the mentally retarded.   There are 30 of
these classes altogether, with a total enrollment of more
than 300.   In 1980, China had 292 schools for the blind
and the deaf-mutes, 9 times that before Liberation.   Total
enrollment was 33,100, or 15 times that before Liberation.

Goal and Length of Study   The school goal for the
blind and the deaf-mutes is to enable them to acquire culture, occupational skill, and communist ethics and to become good, socialist workers.

The length of study for the blind and the deaf-mutes
is far from uniform.   Generally speaking, it is five years
in primary schools and three years in junior middle
schools for the blind; it is eight years in primary schools
and three to four years in junior middle schools for the
deaf-mutes.   Some schools require them to study seven
years in primary school and three years in junior middle
school, for a combined length of ten years.

Curriculum Curriculum in schools for the blind and the deaf-mutes is basically the same as that in ordinary primary and junior middle schools. Primary schools offer such courses as Chinese, mathematics, politics, general knowledge, and physical culture. Junior middle schools offer Chinese, mathematics, politics, physics, history, biology and physiology. Contrary to ordinary schools, schools for the blind and the deaf-mutes offer courses on handicrafts, schools for the blind offer no course on fine arts, and schools for the deaf-mutes offer no course on music.

Chinese language teaching in schools for the deaf-mutes includes learning Chinese ideograms, reading, dialogue, story telling, composition and calligraphy. The students are required to familiarize themselves with the verbal way of correctly expressing themselves and of exchanging ideas with others. They are also to learn to write narratives, simple essays and letters. Through courses on mathematics, they acquire the ability of logical thinking and of solving simple practical problems as found in daily life and production. By taking courses of general knowledge that include natural science, history and geography, they acquire the basic knowledge about nature and social life. Children of low grades are taught to do simple bodily exercise, while those of higher grades are to take regular courses of physical culture, sometimes doing exercise according to radio music. They are also taught to do eye exercise and to protect their eyesight. Through such courses of fine arts as drawing and handicrafts, deaf-mute students acquire the ability for artistic creation.

To help students to be gainfully employed after graduation, schools for the blind and the deaf-mutes offer occupational courses. Schools for the deaf-mutes, for instance, offer courses on carpentry, iron work, embroidery, decorative art, artistic design and sewing. Artistic works created by the class of decorative art in the Taiyuan School for the Deaf-Mutes in Shanxi have been sold abroad. Schools for the blind offer courses on massage and plaiting. To the welcome of the public, many schools for the blind have opened clinics of massage to give treatment to those who need it.

Teaching Plan There is no nationwide scheme in the teaching of the blind and the deaf-mutes. In Beijing, blind children from the first to the fourth grade, by taking 5 to 7 courses, have 28-29 class hours per week. The fifth to the eighth graders, by taking 5 to 8 courses, have 29-30 class hours per week. The fifth and sixth graders perform manual labour for two weeks per year. The length of manual labour per year is four weeks as far as the seventh and eighth graders are concerned. Students below the fifth grade are exempted from manual labour.

By using braille, schools for the blind have been able to use the kind of teaching material normally used in full-time primary and middle schools. In old China, written scripts for the blind were never made uniform; besides, often were dialects used. Beginning in 1953, only the standard spoken Chinese language has been used, and the unified script is often referred to as the new language for the blind.

Teaching material for the deaf-mutes is written, compiled and published by the state. Oral method is used for classroom instruction, supplemented by sign language and finger spelling. One advantage of the oral method is that by watching mouth movement, students learn to know the spelling in the standard spoken Chinese and thus are able to conduct conversation.

Educational Funds and Benefits Expenses for administration and teaching in schools for the blind and the deaf-mutes are larger than those for ordinary primary and middle schools. State funding for a primary school for the blind and the deaf-mutes is as large as that for an ordinary senior middle school located in the same place. State funding for a junior middle school for the blind and the deaf-mutes is as large as that for a secondary normal school located in the same place.

Teachers in schools for the blind and the deaf-mutes are paid a higher salary than those teaching at ordinary schools.

To make sure that students in schools for the blind and the deaf-mutes can concentrate on study, the state provides subsidies for those that have financial difficulties. Blind children and young deaf-mutes with no means of support can live entirely at the state's expense. Expenses of this kind are not included in the quota of subsidies normally granted by the state. Since students are on a part-work and part-study programme, many schools have been either completely or partially self-sufficient in providing room and board for their students.

After graduation, blind and deaf-mute students living in the cities are assigned to work by labour and civil administrative departments of the city government. Those from the countryside would return home to take part in farm work. The blind and the deaf-mutes receive the same pay for the same kind of work.

(8) ADULT EDUCATION

Brief Introduction The purpose of adult education

	Number of Graduates	Number of Students at School	Number of Full-time Teachers
	(in ten thousand persons)		
1) Higher Education	33.58	155.41	3.49
Universities Via Radio and Television	9.27	32.44	0.55
Colleges for Workers and Peasants	6.86	45.53	1.31
Correspondence Colleges	1.54	17.70	0.05
Schools for Middle School Teachers	17.45	77.44	1.63
2) Secondary Education	121.13	804.47	4.32
Secondary Technical Schools for Workers and Peasants	51.97	296.27	1.65
Spare-time Middle Schools	47.97	355.08	1.52
Schools for Primary School Teachers	21.19	153.12	1.15
3) Spare-time Elementary Schools	605.09	1,646.1	6.48
Literacy Classes	538.81	1,220.9	

is to raise the educational level of hundreds of millions of peasants, workers, cadres and other working personnel. The situation regarding the country's adult education in 1980 is shown in the table on p. 96.

Adult education has not progressed to the extent as the present situation demands. A great number of young and middle-aged people remain illiterate because of the "cultural revolution"; 5 per cent of workers and about 30 per cent of young and middle-aged peasants are illiterate. Many people educated during the "cultural revolution" are not really educated. The Ministry of Education has proposed that illiteracy must be wiped out among the young and middle-aged workers in two or three years and that, wherever local conditions permit, illiteracy among young and middle-aged peasants must be wiped out in three to five years. It is further proposed that before the year of 1985 all young workers and staff members in factories and mines must attain an educational level higher than that of a junior middle school graduate. In the meantime, to meet the urgent demand of the state for professional personnel and the demand of unemployed young people for further study, education by TV evening college and correspondence education must be further developed. There must be more spare-time middle schools and colleges to enroll government cadres, working staff from industrial establishments, and young people waiting for job assignments.

Education for Peasants At present the emphasis is on the education for young and middle-aged peasants and rural cadres at the grass-roots level. Their education take many forms, such as peasant literacy classes, literacy classes for peasants in slack seasons, part-farming and part-study classes, and spare-time study classes in production teams. In addition, there are spare-time primary and junior middle schools and technical schools for peasants under the aegis of production brigades and people's communes. These schools offer courses of junior middle schools and also courses of basic knowledge on agricultural science. In some areas correspondence courses for educated youth are also available on a trial basis.

As education for peasants requires a great number of teachers, teachers must come from local sources and "those who know must teach". Most teachers are educated youth in the countryside but full-time teachers from nearby primary and middle schools and technicians from agronomical and livestock-breeding stations are also invited to help on a part-time basis. These part-time teachers will be compensated for their absence from their regular jobs according to local standards.

Education for Workers and Staff   This kind of education means, principally, spare-time and full-time training of workers and staff members in science, technology, business management and general knowledge.   First, leading cadres of factories and mines, plus administrative personnel and technicians, can study the method of scientific management and other professional and technical knowhow on a full or part-time basis.   Second, young workers and staff members employed since the "cultural revolution" could be enrolled in spare-time classes of regular secondary school and secondary technical school level, so eventually they would acquire the academic qualifications of a junior middle school graduate. Those who have attained the educational level of a junior middle school graduate should of course strive to reach a senior middle school graduate's level. Third, there should be more spare-time universities and universities for workers and staff members. In 1979, 72 universities offered correspondence courses, including 30 undergraduate subjects and 17 subjects of a technical nature. The same year, there were 200 colleges for workers and staff members in Shanghai. Total enrollment was as large as 50,000. Students are accepted after succeeding in a city-wide entrance examination, and they study on a full-time, part-time, or spare-time basis. More than 100 different concentrations are offered.

Elementary classes are generally conducted by individual factories and workshops that, jointly, may offer advanced classes. Large enterprises, individually or jointly, may run their own spare-time universities for workers and staff members. Others may be run by specialized companies or bureaus, academic societies or associations. The government's educational departments may operate workers' schools or educational programmes via radio and TV, while regular colleges educate more by means of correspondence or night university.

The enterprises select their own teachers for educating their workers and staff, supplemented by college graduates assigned by the state. Engineers, technicians and teachers from regular schools may also be invited to teach, to be remunerated according to the number of class hours they teach. Advanced institutes for teachers and educational institutions of higher learning may wish to take the responsibility of training teachers for schools of workers and staff members.

Education by Radio and TV In China, education by radio and TV is a new form of education for adults. In addition to the educational programmes by the central TV university, similar programmes are offered in 28 provinces and municipalities. Founded early in 1979, the

TV education today has an enrollment of 324,400 students taught by a teaching staff of 5,500.

TV universities have as their students workers and staff members who have attained the educational level of a senior middle school graduate. Once approved by their own units, workers, school teachers, army officers and soldiers, and also educated youth in cities and the countryside can enroll once they pass an entrance examination.

Among the subjects of study offered by TV universities at present are basic general-purpose courses of science and technology and special courses on higher mathematics, physics, chemistry, engineering mechanics, fundamentals of electrical engineering, mechanical design, electronic technology, fundamentals of computer and computer language, biology and English. It takes the central TV university three years to complete the transmission of each of the courses offered. Specialized courses, basic technical courses and courses on politics are generally offered by local TV universities.

As supplementary reading material has been compiled and issued in advance, the greatest advantage of teaching by TV is the largeness of the territory and the length of the distance it can cover. The TV teaching is invariably accompanied by face-to-face coaching, experiments, and practical work.

The TV universities have instituted a complete administrative system where the upper-layer work is closely coordinated with that below. Under the leadership of the Ministry of Education and the Broadcasting Administrative Bureau, they see to it that a unified teaching plan be worked out, teachers invited and teaching materials compiled and distributed. They are responsible for recruitment of students via entrance examinations and make questions for final examinations at the end of each term. Local TV universities, on the other hand, are under the leadership of the educational departments and broadcasting administrative bureaus in various provinces, municipalities and autonomous regions. Under them are branch schools or working stations that are entrusted with the work of recruiting full-time or part-time instructors arranging laboratory work, providing teaching facilities, equipment and funds, and other administrative work. The study class is the basic unit of the TV universities, supervised by a full-time or part-time instructor and an administrative cadre. Each study class has a class committee and elects a class representative who is in charge of attendance, finance, sports and recreational activities, laboratory work and coaching, and other administrative work.

TV universities have three kinds of students, i.e., full-course, dual-course and single-course, studying respectively on a full-time, part-time and spare-time basis.

Correspondence School and Evening University Correspondence school and evening university are an important form of spare-time education that made good progress in the fifties and sixties. Now they are being restored. In 1980, 69 regular universities undertook education by correspondence, and 24 regular universities and colleges operated evening universities. Total enrollment was 189,000. Prospective students of correspondence school and evening university are employed staff members and workers who have had the academic level of a senior middle school student. Once enrolled, they are taught by the same teachers who teach the regular students and are provided with the same facilities and

equipment. In other words, the quality of their education is assured. Being flexible, education by correspondence is most welcome among the masses. Apart
from correspondence education made available by institutions of higher learning, there are also a number of
independent correspondence institutes. An outstanding
example is the ' posts and telecommunications correspondence school that is operated on a nationwide scale.
Its centre is the Beijing Correspondence Institute of Posts
and Telecommunications Education, formerly correspondence education department of the Beijing Institute
of Posts and Telecommunications. It offered four subjects
of study and had at one time a student body of more than
1,100. In 1980 it planned to add 3,000 more students. It
has more than 30 correspondence branches and 95 correspondence stations across the country, jointly forming
a nationwide network. Society as a whole has now paid
great attention to correspondence school and evening university. In 1980 the Ministry of Education called a meeting on correspondence school and evening university,
which was attended by representatives of institutions of
higher learning. It planned to develop further an education of this kind.

**Pay and Other Benefits for Workers and Staff
Members Studying at School** Workers and staff members
who have received permission from their leaders to study
in TV universities or spare-time schools will be paid their
normal wage or salary. Full-time students completely
released from regular work will retain their non-produc-
tive welfare benefits but will stop receiving productive
bonuses and productive benefits for labour protection.
Students studying on a part-time basis are to receive one
half of the productive bonuses and benefits for labour

protection. As for those who study during their spare time, all benefits will remain the same.

Arrangement of Work for Graduates A system of examination and assessment is strictly enforced in all study classes and spare-time schools. After completing a required number of courses and by passing successfully the required examinations, students are entitled to graduation. The record of their attendance and accomplishment would be formally admitted as the chief criterion for their work assignment within their respective units, salary increase or promotion, though the state is not obligated to assign them to work in units other than their own. Workers and staff members, after graduation, would return to their former posts. When necessary, changes would be made in their work so as to take advantage of what they have learnt. Outstanding workers and peasants who have done exceptionally well in their study may be promoted as cadres or technicians. As for educated youth waiting for work assignments, the schools, though not responsible for giving them jobs, will recommend them to proper organizations or agencies that might employ them.

(9) EDUCATION FOR FOREIGN STUDENTS IN CHINA

Beginning in 1950, China has accepted foreign students to study in China. The exception occurred during the "cultural revolution" when foreign students were not accepted. In 1973, China once again accepted foreign students to study in China. From 1973 to 1979, altogether 2,473 foreign students from 76 countries and regions had come to study in China. During the past two years, with the development of economic and cultural exchanges between China and foreign countries, not only the number of foreign students increased but the subjects they studied also multiplied. At the end of 1980, there were 1,382 foreign students in China, majoring in 47 subjects in 42 institutions of higher learning in 12 provinces, municipalities and autonomous regions. An accomplishment of this kind, though great, is anything but adequate in meeting the demand for student exchange. However, owing to the fact that China at present is limited economically, it can only do what it can in the hope that improvement will be continually made as conditions permit.

5. TRAINING OF TEACHERS

(1) TEACHING FACULTY IN THE INSTITUTIONS OF HIGHER LEARNING

With the development of higher education since the founding of the People's Republic, there has been a speedy increase of college faculty. The number of college teachers increased from 16,000 in 1949 to 246,900 in 1980, of whom 4,212 were professors, 13,788 were associate professors and 119,100 were lecturers.

Members of college faculty are drawn mainly from two sources: outstanding graduates from one's own university or college or postgraduates from other institutions of higher learning. In the last few years, professional training for in-service teachers has been strengthened to compensate the inadequate preparation caused by the "cultural revolution" among young and middle-aged

Colleges and Universities with Foreign Students and the Subjects of Study They Offered, 1980

Name	Subjects	Student Type	Length of Study
Beijing Languages Institute (Beijing)	modern Chinese, basic Chinese	undergraduate students students for advanced study	four years one year
Beijing University (Beijing)	history of China, Chinese, ancient Chinese history, modern Chinese history, archaeology, political economy, Chinese literature, philosophy	undergraduate students students for advanced study	four years one year
Fudan University (Shanghai)	classical Chinese literature, modern Chinese literature, ancient Chinese history, modern Chinese history, history of Chinese philosophy	students for advanced study	one year
Nanjing University (Nanjing)	Chinese, Chinese literature, history of China, history of Chinese philosophy, political economy	students for advanced study	one year
Nankai University (Tianjin)	Chinese, Chinese literature, ancient Chinese history, modern Chinese history, contemporary Chinese history	students for advanced study	one year

Name	Subjects	Student Type	Length of Study
Liaoning University (Shenyang)	Chinese, classical Chinese literature, modern Chinese literature, ancient Chinese history, modern Chinese history, contemporary Chinese history	students for advanced study	one year
Shandong University (Jinan)	ancient Chinese, modern Chinese, classical Chinese literature, modern Chinese literature, ancient Chinese history, modern Chinese history	students for advanced study	one year
Qinghua University (Beijing)	industrial automation, electronic computer, machine-building technology and equipment, hydraulic engineering, architecture, industrial and civil construction, power plant and electric power system	undergraduate students	five years
Tianjin University (Tianjin)	radio technology, industrial electric automation, power plant and electric power system, electric machinery,	undergraduate students	four years

Name	Subjects	Student Type	Length of Study
Tianjin University (Tianjin)	water conservancy construction, machine-building technology and equipment	undergraduate students	four years
Tongji University (Shanghai)	industrial and civil construction, architecture	undergraduate students	four years
Nanjing Engineering Institute (Nanjing)	radio technology, highway engineering, machine-building technology and equipment, power plant and electric power system	undergraduate students	four years
Beijing Institute of Posts and Tele-communications (Beijing)	radio technology (posts and tele-communications)	undergraduate students	four years
Beijing Iron and Steel Engineering Institute (Beijing)	mining, electric metallurgy	undergraduate students	four years

Name	Subjects	Student Type	Length of Study
Jiaotong University in North China (Beijing)	diesel locomotives, railway rolling stock, railway signals, railway engineering, railway transport	undergraduate students	four years
Shanghai Chemical Engineering Institute (Shanghai)	chemical engineering, basic organic chemical engineering, polymer chemical engineering, oil refining	undergraduate students	four years
Shanghai Engineering Institute (Shanghai)	electrical engineering, machinery design and technology	undergraduate students	four years
Shanghai Textile Engineering Institute (Shanghai)	cotton textile, woollen textile, machine weaving, dyeing and finishing, textile machinery, chemical fibres	undergraduate students	four years
East China Water Conservancy Institute (Nanjing)	farm water conservancy	undergraduate students	four years
Wuxi Light Industry Institute (Wuxi)	food processing	undergraduate students	four years

Name	Subjects	Student Type	Length of Study
Jilin Engineering Institute (Changchun)	agricultural machinery	undergraduate students	four years
Changchun Geological Institute (Changchun)	general and mine survey	undergraduate students	four years
Xian Highway Engineering Institute (Xian)	use and maintenance of motor vehicles, highway engineering, bridge and tunnel	undergraduate students	four years
Zhejiang Agricultural University (Hangzhou)	agronomy, soil chemistry	undergraduate students	four years
Guangxi Agricultural Institute (Nanning)	agronomy, animal husbandry and veterinary science	undergraduate students	four years
Beijing Medical College (Beijing)	medical science	undergraduate students students for advanced study	five years one year
China Medical University (Shenyang)	medical science	undergraduate students students for advanced study	five years one year

Name	Subjects	Student Type	Length of Study
Shanghai No. 1 Medical College (Shanghai)	medical science	undergraduate students	five years
		students for advanced study	one year
Shanghai No. 2 Medical College (Shanghai)	medical science	undergraduate students	five years
		students for advanced study	one year
Zhongshan Medical College (Guangzhou)	medical science	undergraduate students	five years
		students for advanced study	one year
Guangxi Medical College (Nanning)	medical science	undergraduate students	five years
Beijing Traditional Chinese Medical Institute (Beijing)	traditional Chinese medical science	undergraduate students	five years
		students for advanced study	one year
Guangzhou Traditional Chinese Medical Institute (Guangzhou)	traditional Chinese medical science	undergraduate students	five years
		students for advanced study	one year

Name	Subjects	Student Type	Length of Study
Nanjing Pharmaceutical Institute (Nanjing)	pharmacology	undergraduate students	four years
		students for advanced study	one year
Central Academy of Fine Arts (Beijing)	traditional Chinese painting, history of Chinese fine arts	students for advanced study	one year
Zhejiang Academy of Fine Arts (Hangzhou)	traditional Chinese painting	students for advanced study	one year

teachers. To raise the professional level of teachers and to enable them to learn from each other, the Ministry of Education sponsors forums on holidays where college teachers listen to lectures or reports given by Chinese and foreign experts. During the summer of 1980, the Ministry of Education sponsored 16 seminars on basic microorganism, solid state physics, genetics, operational research, management system, quantum chemistry, theory on the structure of atom and molecule and other subjects, that were attended by more than 2,000 teachers.

In recent years, institutions of higher learning have paid particular attention to the training of middle-aged teachers in order to raise their professional level. To create as quickly as possible a large teaching faculty in the field of natural and social sciences and to change the present state of affairs in which there is a serious shortage of professional personnel needed by the state, they have done everything possible to enhance the academic quality of college faculties, such as encouraging them to enroll for advanced study or to do research within a fixed period of time. Meanwhile, they are enlarging the enrollment of postgraduates to assure a continuous outflow of new qualified teachers for institutions of higher learning.

(2) TEACHING FACULTY OF PRIMARY AND MIDDLE SCHOOLS

Teachers on the level of primary and middle schools are mostly graduates from normal schools or colleges. According to the requirement of the state, teachers of senior middle schools must be graduates from normal colleges or have an educational level equivalent to that of a university graduate; teachers of junior middle schools must be of or above the level of a technical college graduate; and teachers of primary schools must be of or above the level of a secondary normal school graduate.

Since the founding of the People's Republic, normal education has made great progress, and it has provided the state with large numbers of teachers. During the period from 1949 to 1979, 667,567 persons graduated from the nation's normal colleges and universities, constituting 22 per cent of the nation's graduates from institutions of higher learning. Graduates from secondary normal schools numbered 2,474,779, or 46 per cent of the total number of graduates from secondary specialized schools. In 1979, China had 161 normal colleges and universities, or 25 per cent of the total number of institutions of higher learning in the country. Total enrollment in normal colleges and universities was 310,000, or more than 30 per cent of the total number of college and university students, second in size only to those who studied engineering. The enrollment was 25 times that of 1949, the largest increase among the various disciplines. There were then 1,053 secondary normal schools with an enrollment of 480,000, 7.8 times that of 1949. Yet they cannot satisfy the need of the state for teachers because of the speedy development of primary and middle school education. At present, there are still teachers in primary and middle schools whose educational level cannot meet the requirements of the state.

To solve the problems as mentioned above, educational agencies have adopted the following measures:

A. They strive to do a good job in normal education to guarantee the supply of qualified teachers for primary and middle schools.

The task for normal colleges and universities is to train teachers for middle schools. The policy at present is to improve the quality of education offered by the existing teacher training, raise teaching standards and make sure that graduates be not only politically and morally superior but also professionally proficient in the basic knowledge of science and culture, besides being thoroughly familiar with educational theory and practice. Only when quality is assured would there be enlargement of enrollment and the establishment of new normal colleges and universities. In 1980, preparations were made to establish four normal colleges. Recently the State Council has given approval for the establishment of 11 more. The length of study for normal colleges and universities is four years, while that for technical colleges is only three years.

By enrolling junior middle school graduates, secondary normal schools train teachers for primary schools, and the length of study is three years. Some favourably endowed secondary normal schools are even entrusted with the task of training in-service teachers from primary schools.

Students studying in normal colleges or schools are all provided with scholarships by the state. To guarantee the stability of the teaching force, in-service teachers of primary and middle schools are only allowed in principle to take the entrance examination given by normal colleges.

B. Educational agencies are doing their best in strengthening the training of in-service teachers.

In-service teachers of primary and middle schools, who wish to pursue advanced studies, are, in principle, encouraged to study by using their spare time. The study can take any of the following forms:

(a) Their own schools train them in the sense that the old teach the young, they teach one another, or they prepare lessons and do research collectively. The ideal is to combine teaching with one's own study.

(b) Experienced teachers are organized to give coaching or lectures in different places or at special points.

(c) Training courses varying in length of time are offered for teachers.

(d) Correspondence school and radio or TV education are utilized to raise teachers' professional level.

Colleges and schools for advanced study take upon themselves the task of training teachers. In various provinces, municipalities and autonomous regions there are normal schools and colleges, colleges of education, and correspondence colleges. At the prefectural, city and county levels there are colleges of education and training schools for teachers. People's communes have also courses of advanced study that are attended by teachers for their own advantage. Thus there is a nationwide network of teacher training. In 1979 there were 30 colleges for advanced study and correspondence colleges at the level of provinces, municipalities and autonomous regions where teachers pursued their studies. At the level of prefectures, cities, and counties, there were more than 2,000 colleges of education and teacher training centres that served the same purpose. In addition, more than 50 normal colleges and universities operated correspondence schools and training classes for middle school teachers. Across the country more than 800,000 middle school teachers took part in spare-time study, accounting for 33 per cent of the teachers who needed advanced study. The number of primary school teachers taking part in spare-time study totalled 1,300,000, accounting for 47 per cent of the teachers who were in need of further study. Apart from those taking part in short-term training, there were even a great number of teachers taking courses for systematic study.

Teachers are assured of adequate time for spare-time study. Generally speaking, teachers have half a day off every week for professional study. Having passed required examinations, primary and middle school teachers with qualifications equivalent to those of college or secondary normal school graduates would be granted corresponding diplomas, and their new academic status would be recognized by the state as bona fide.

6. ADMINISTRATIVE STRUCTURE AND LEADERSHIP SYSTEM IN EDUCATION

(1) CENTRAL ADMINISTRATIVE STRUCTURE

The Ministry of Education is the highest administrative organ of education in China, under the direct jurisdiction of the State Council. In observance of the educational policy of the state and decisions of the State Council, it formulates and issues regulations governing the work of primary and middle schools as well as institutions of higher learning. It plans educational development, finance, personnel and teaching programmes, including those on students' productive labour and practical training. It initiates laws and directives that govern educational work for the whole country.

The Ministry of Education has under it administrative organs to carry out its assigned duties. Under it are also 35 key colleges and universities, the People's Educational Publishing House, the *People's Education* magazine, the Research Institute of Educational Science and the Central Open University by Radio and TV.

(2) LOCAL ADMINISTRATIVE ORGANS

All provinces, autonomous regions and centrally administered municipalities have within their respective jurisdictions educational bureaus. Under the unified leadership of the local people's government, these bureaus carry out the administrative work of education in their own areas in line with the directives or regulations issued by the Ministry of Education. Under the jurisdiction of people's governments in various counties, autonomous prefectures, cities and municipal districts are educational departments. In every people's commune and town there is likewise an educational section to be in charge of the administrative work of local education on behalf of the educational departments of counties or autonomous prefectures. Led directly by people's governments of the corresponding level, local educational administrative departments also work under the guidance of the educational authorities at the next higher level.

(3) LEADERSHIP SYSTEM IN SCHOOLS

A system of dual leadership is adopted in the administration of the institutions of higher learning. The Ministry of Education and educational bureaus in various provinces, municipalities and autonomous regions provide leadership over the administration of comprehensive universities, poly-technic universities and normal universities and colleges. With the exception of a small number of key colleges and universities which are administered chiefly by the Ministry of Education, most of the institutions of higher learning are administered chiefly by local educational departments. Single-discipline colleges and institutes specializing in engineering, agriculture, forestry, medical science, business administration and economics, arts and physical culture function mainly under the administration of relevant ministries and of the State Council or responsible departments in various provinces, municipalities and autonomous regions. Take the Beijing Iron and Steel Engineering Institute and the Central Academy of Fine Arts as an example. They are respectively under the jurisdiction of the Ministry of Metallurgy and the Ministry of Culture. The Ministry of Education only makes it sure that guidance and instructions be provided when it comes to problems related to policy and regulations as common in education.

Secondary specialized schools function under the jurisdiction of either the ministries of the State Council or local authorities. Among the schools under the ministries of the State Council are those under the direct leadership of these ministries' enterprises or institutions. Local secondary specialized schools are the responsibility of educational departments in various provinces, regions and municipalities. The Ministry of Education is only responsible for a school's development plan, recruitment plan, teaching plan, and teaching material of ordinary courses. The teaching plan of specialized courses, the teaching material of specialized courses, the personnel and its funding will be the responsibility of that ministry of the State Council or that of local authority under which the school operates.

Ordinary primary and middle schools in medium-sized or large cities operate under the leadership of district educational departments. Those in the countryside and small towns operate under the leadership of either the respective counties' educational departments or the management committees of the people's communes or the educational agencies of town governments. Primary and middle schools funded by factories, mines and institutions are respectively controlled by the funding units.

(4) LEADERSHIP SYSTEM WITHIN VARIOUS EDUCATIONAL ESTABLISHMENTS

All educational institutions are run by a responsibility system of division of labour under the presidents or principals, who in turn are responsible to the Party organizations of the institutions concerned. Serving as a chief administrator, a president or principal shoulders the overall responsibility at school. Important matters have to be submitted to the school's Party committee for deliberation; once a decision is reached, the president or principal has the responsibility to carry it out. Major problems of a daily routine nature will be discussed and resolved by school committees composed of leading cadres from various departments and representatives of teachers. Full faculty meetings or representative conferences are regularly held, during which both suggestions and criticisms are made in connection with the work of the leadership.

7. SCIENTIFIC RESEARCH IN EDUCATION

(1) THE TASKS AND PRINCIPLES OF SCIENTIFIC RESEARCH IN EDUCATION

Under the guidance of Marxism-Leninism and Mao Zedong Thought, the purpose of scientific research in education is to study, principally, the country's practical problems involving education. Another purpose is to study the theory, history and present status of education at home and abroad, so we shall be well informed on the law governing the development of education, so that we can formulate our own educational policy, reform our educational system, and improve the teaching material and method we use. Gradually we shall be able to establish a scientific system of socialist education in China.

The policy in guiding educational research is, first, to "make the past serve the present and foreign things serve China". All the fine things from the past, whether foreign or domestic, should be critically absorbed to improve the present education. The second guideline is to carry out thoroughly the idea: "Let a hundred flowers blossom and a hundred schools of thought contend". In other words, we shall have not only free discussion of academic problems but also all kinds of experiments, comparisons and research for the introduction of educational reforms. The third guideline is to recognize reality which, when combined with theoretical study and present need, is the key to solving present educational problems.

(2) RESEARCH INSTITUTIONS IN EDUCATION

The Central Institute of Educational Science and the Chinese Educational Society are the national research institutions in education.

The Central Institute of Educational Science This is an institution directly under the leadership of the Ministry of Education. In professional matters it is guided by the Chinese Academy of Social Sciences.

The institute has a planning group responsible for the co-ordination of efforts in planning the country's scientific research in education, besides the routine work in carrying out the planning. The institute has nine departments doing research on educational theory, methodology, modern educational technique, educational psycology, history of education, educational system, information on education, school management and pre-school education. In addition, there is an academic committee composed of senior research personnel from various academic departments to discuss, examine and evaluate the result of each research project. It offers opinions on the evaluation and promotion of research and associate research fellows, on the invitation and appointment of visiting researchers and correspondence researchers, and on the granting of academic titles or degrees. It sponsors academic forums and promotes academic exchanges both at home and abroad.

The titles for educational research personnel include research fellow, associate research fellow, assistant researcher and research apprentice, graded according to their professional abilities and academic achievement. The institute also invites outside experts as special researchers or correspondence researchers to take part in planned projects or projects of their own choice.

The institute has its own laboratory and library, in addition to a publishing house and an experimental ground of education. Its publications include *Educational Study* and *Education in Foreign Countries*.

The Chinese Educational Society This is an academic society of educational workers. Its aim is to unite scientific research personnel in education, teachers, and administrators in carrying out the activities and popularization of educational research so as to help the country in developing scientific research in education.

The Chinese Educational Society has under its jurisdiction nine nationwide professional research societies: the Research Society of Educational Science, the Research

Society of the History of Education, the Research Societ
of Foreign Education, the Research Society of Marxis
Education, the Research Society of Chinese Teaching i
Middle Schools, the Research Society of Chinese Teachin
in Primary Schools, the Research Society of Chinese
Teaching Methodology, the Research Society of Pre-schoo
Education and the Society for the Work of Young Pioneers
Altogether they have a membership of more than 3,500
Professional research societies being planned include th
Research Society for Foreign Languages Teaching, th
Research Society of Mathematics Teaching in Primary anc
Middle Schools and the Research Society of Special Edu-
cation. Frequently these research societies sponsor aca-
demic exchanges, including the issuance of academic
reports, the holding of annual academic meetings, and the
publication of periodicals and research papers.

Now 28 provinces, municipalities and autonomou
regions have either established or restored educationa
societies, the membership of which totals more thar
20,000. Even some prefectures and cities have educationa
societies.

(3) PERSONNEL FOR EDUCATIONAL RESEARCH

Owing to the devastating effect caused by the "cultural revolution", only a small number of veteran researchers has remained, and many of them are already
advanced in years. Consequently, there is a shortage of
personnel to carry on educational research. Efforts are
now being made to train postgraduates and to increase
this professional force. An important source of this professional force comes from normal colleges and universi-

es: teachers, postgraduates and upper-grade students,
nce they are widely distributed geographically, besides
eing well grounded in professional theories. Another
ource of educational research is the broad masses of
eachers in primary and middle schools, who are rich in
xperience and can provide favourable conditions for edu-
ational experiments.

Chapter Two

NATURAL SCIENCES

1. SCIENTIFIC AND TECHNOLOGICAL ACHIEVEMENTS IN ANCIENT TIMES

For several thousand years the Chinese people of all nationalities have invented and created and made great contributions to the development of science and technology.

Astronomy, mathematics, agronomy and medicine were the sciences in which China excelled in ancient times. The silk technology, made world famous because of the Silk Road; the manufacture of porcelainware, so esteemed that the country itself acquired the name of China; the paper-making, printing, gunpowder, compass, all well-known both at home and abroad; and the ancient structures, such as the Great Wall, and the imperial palaces and garden buildings in Beijing — all this can be viewed as true representatives of China's achievements in science and technology in ancient times. Joseph Needham, a famous British historian of science, once said that ancient China's achievements in science and technology were so far ahead during the period between the 3rd and 13th century that the West lagged too far behind to catch up.

Astronomy In order to arrange farm and animal husbandry activities in accordance with the change of seasons, ancient people badly needed astronomical knowledge in making calendar. Ancient Chinese made great contributions not only to calendar making, but also to the keeping of astronomical records and the manufacturing of astronomical instruments.

In Chinese history more than 50 calendars were promulgated at one time or another. The calendar made in about the 5th century B.C. had 365.25 days for a year. Subsequently the number of days for a year was made more and more correct. In the 5th century a famous scientist named Zu Chongzhi (429-500) came out with 365.2428 days for a year. In 1199 the astronomer Yang Zhongfu made further improvement with 365.2425 days for a year — the same yearly length as it was used in the Gregorian calendar. But the Chinese discovery appeared more than 400 years earlier. Forecasting solar and lunar eclipses was one of the main purposes in the making of ancient calendar, and it required precise knowledge of the laws that governed the movement of the sun and the moon. Ancient Chinese accumulated astronomical data through the use of continually improved observation instruments, and they, as a result, became more and more precise in locating the relative positions of the sun and the moon. Besides, they used then the world's advanced mathematical methods, especially the interpolation method, which had been discovered by China at that time, in tracing the orbits of the sun and the moon. Astronomical constants were calculated more and more precisely with the passage of time.

Another important achievement in ancient astronomy consists of the rich records of celestial phenomena, among which the records of sunspots, comets, polar lights, novas and meteorite showers are not only the earliest and

continually made in the world but also rich in conten
From the remote antiquity to the 18th century, for in
stance, China recorded more than 90 novas and super
novas, which have attracted great attention among moder
radio astronomers.

Another outstanding achievement in China's ancien
astronomy is the manufacture of astronomical instru
ments, such as the armillary sphere, which was used t
locate celestial bodies, and the celestial globe, whicl
demonstrated celestial phenomena. In the 11th century
the Song scientist Su Song supervised the construction of ;
12-metre-high instrument, which could give correct tim
of the day besides serving as an armillary sphere. I
was the oldest astronomical clock, employing the constan
flow of water to rotate water wheels at intervals so as t
drive the machine. Ancient Chinese astronomers also use(
continually improved instruments to scan the heaven
and draw all kinds of star maps and charts. Amon;
them the Shi Shen Star Catalogue compiled in about th
4th century B.C. was probably the world's earliest or
record. The Dunhuang Star Chart (now preserved in th
British Museum) of the early Tang Dynasty (about the 7th
or 8th century) and the stone star chart carved in Suzhou
in the Southern Song Dynasty (between the 12th anc
13th century) are also world-famous. The latter has or
it more than 1,400 stars.

Mathematics China was the first country tha
adopted the decimal system. In approximately the 4th
century B.C., the Chinese were proficient in the use of the
decimal system and the fractions for the four arithmetica
operations. The discovery of the decimal system is saic
to be only second in importance to the adoption of fire
for the advance of human civilization.

Jiu Zhang Suan Shu (*Nine Chapters on the Mathematical Art*), completed in the 1st century, is not only a Chinese classic of mathematics but also a well-known scientific work in the world. Like Euclid's *Elements* which served as a definitive textbook in the West, the Chinese work did the same for those who studied mathematics in ancient China. In this book appear negative numbers and their computation, plus solutions for simultaneous linear equations. It was one of the most advanced works on mathematics at that time.

In the Southern and Northern Dynasties (420-589), a mathematical genius named Zu Chongzhi (429-500) calculated the true value of π between 3.1415926 and 3.1415927. At the same time he pointed out that the fraction value of π was $\frac{355}{113}$. Only a thousand years later was Zu's achievement equalled by European mathematicians.

During the Song-Yuan period, between the 13th and 14th century, Chinese mathematics underwent a new development. Famous mathematicians emerged, such as Qin Jiushao, Li Ye, Yang Hui and Zhu Shijie. New mathematical heights were attained.

By the middle of the 13th century or earlier, Chinese mathematicians had succeeded in working out the root of equations of higher degree by numerical solution, and the method is explained in detail in *Shu Shu Jiu Zhang* (*Mathematical Treatise in Nine Sections*), written by Qin Jiushao in 1247. This method of Qin's is equal to Horner's in the West, but it appeared nearly 600 years earlier. Moreover, simultaneous multivariate higher algebraic equations of the fourth order or below are solved in Zhu Shijie's mathematical works published in 1303. Because of his achievements, Savton, a famous American historian of science, considers Zhu Shijie the world's most outstanding mathematician in Middle Ages. Besides, the research work done by Chinese mathematicians on linear congruences, higher order arithmetic series and the rule of double false positions was not matched by Western mathematicians until several hundred years later.

By the 16th century the Chinese abacus had spread far and wide across China. The fact that this simple and convenient calculating instrument is still used in China seems to indicate that it was the best calculator before the invention of the modern computer.

Agronomy   China is one of the centres where crops began. It produces many kinds of crops with even more specimens. In the Zhou Dynasty, over 3,000 years ago, there were such crops as glutinous millet, broomcorn millet, millet, rice, barley, wheat, soya bean and bast fibre. The agricultural book of *Qi Min Yao Shu* (*Important Arts for the People's Welfare*), written in the 6th century, records over 86 specimens of millet, indicating that Chinese agriculture had reached a very high level of development at that period. During the last several hundred years, local gazettes in different parts of China record numerous specimens of various crops.

Intensive cultivation, starting in the 5th century B.C. or earlier and being improved afterwards, became a fine tradition of ancient Chinese farming. Full utilization of soil and its improvement, combination of soil use with its maintenance, preservation of soil moisture, application of fertilizer, seed selection, meticulous field management, etc. — all this formed parts of a complete system in farming theory and practice. A foreign scholar, when commenting on China's intensive cultivation, says that it is not a coincidence that China, even from a global view-point, had a truly outstanding system of farm theory as early as the 6th century.

Having a long history, China has accumulated rich knowledge of agronomy and farming technology. Long ago, a number of agricultural works appeared, reflecting the achievements of ancient Chinese in their successful struggle against natural environment. The earliest extant agricultural treatises that have been handed down from one generation to the next were written more than 2,000 years ago. According to an incomplete listing, ancient Chinese books on agriculture number more than 370, including those that have been long lost. Among them the better known are *Qi Min Yao Shu* (*Important Arts for the People's Welfare*) by Jia Sixie, the 12th century's *Chen Fu Nong Shu* (*Agricultural Treatise of Chen Fu*), the 13th century's *Wang Zhen Nong Shu* (*Agricultural Treatise of Wang Zhen*) and the 17th century's *Xu Guangqi Nong Shu* (*Agricultural Treatise of Xu Guangqi*).

Medicine (See *Sports and Public Health* of the "China Handbook Series")

Technological Invention *Metallurgy* We must first mention the metallurgy of bronze. Most of the fine bronzes made in ancient China are ceremonial vessels, musical instruments, weapons and daily-use utensils. The largest bronze excavated so far is the rectangular Si Mu Wu Ding cauldron that is 133 cm high, 110 cm long and 78 cm wide and weighs 875 kilogrammes. The bronze artifacts excavated recently from Grave B of Marquis Zeng of the 4th century B.C. in Suixian County, Hubei Province, total about 10 tons in weight. Among them was a set of 64 bells weighing 2.5 tons. The high technology necessary to the making of these bronze artifacts indicates that the Chinese people at that time had already mastered not only the principle of casting by stages to achieve complex configurations but also the welding technology and probably even the wax-replacement method as well. The chime of bells speaks for the high level of acoustics and musical theory in ancient China.

Between the 4th and 5th century B.C., the "six compositions", six rules of compounding proportions of bronze and tin in bronze, were summarized and recorded in the official handicraft handbook entitled *Kao Gong Ji (Artificers' Record)*. They are the principles of making alloys which the Chinese discovered at that time.

Between the 5th and 6th century B.C., China mastered the technology of smelting pig iron and "block iron". (In early China, iron smelting was done at temperatures of 800°C-1,000°C, the ores being directly reduced with charcoal. The wrought iron obtained by this method was called "block iron". It was spongy when it emerged from the furnace. The earliest large-size object cast from iron can be found in Cangzhou, Hebei Province, the iron lion cast in A.D. 953 which weighs more than 50,000 kilogrammes. At the time when the Chinese mastered the technology of iron smelting, the technique of making steel by repeatedly forging carburized iron blocks and of softening pig iron was also invented. Later, they learned to make puddling steel and co-fusion steel. The process of making co-fusion steel was first smelt pig iron of high carbon content and then pour it on wrought iron to carburize it. It became steel after being tempered in quenching liquid. The method of making co-fusion steel had been a most advanced technique before the modern crucible steel-smelting was invented. For more than a thousand years China's iron and steel metallurgy and output were the best in the world.

Besides, China was the first country in the world to smelt zinc. In or before the 16th century, China began zinc-smelting and later passed on the technique to Europe.

Architecture Ancient Chinese architecture formed an independent school of its own. The Great Wall erected in the 2nd or 3rd century B.C. and rebuilt continually afterwards has remained an architectural marvel of the world. It is one of the two man-make structures that can be seen from satellites.

The architectural form of Chinese imperial palaces is a combination of platform and wooden structure. The palaces erected in various dynasties, especially those erected at the beginning of the 15th century during the Ming Dynasty, are grand on scale and solemn in style, with complex structure and logical layout. *Rules of Architecture*, written by Li Jie in 1103, is a standard text and a valuable document in the architectural history of China and the world.

Religious buildings represent another achievement in the history of Chinese architecture, especially ancient pagodas with many storeys that still attract the attention of both Chinese and foreign tourists. The earliest brick pagoda that has survived is the Songyu Temple Pagoda erected early in the 6th century and located in Dengfeng County of Henan Province. The earliest wooden pagoda is the Fugong Temple Pagoda erected in 1056 and located in Yingxian County of Shanxi Province. And the earliest stone pagodas are the two granite pagodas erected between 1228 and 1250 and located inside the Kaiyuan Temple of Quanzhou County, Fujian Province.

Chinese cities, especially those that had served as the nation's capitals, were constructed on a grand scale. Changan and Luoyang during the Han and Tang periods and Beijing during the Yuan, Ming and Qing periods were the world's largest cities.

Ancient China was equally outstanding in the field of bridge building, and a number of famous bridges were erected then. For instance, the Anji Bridge (also called Zhaozhou Bridge) at Zhaoxian County in Hebei Province is the earliest stone arch bridge erected in about 607. Its total length is 50.82 metres, and its arch span is 37.02 metres. The arch, which provides a gradualness of the bridge surface, is different from the traditional semicircle. It helps create the world's first spandrel bridge. Another bridge worthy of mention is a large stone beam bridge, called Wanan Bridge (or Luoyang Bridge), located at Quanzhou of Fujian Province. Its construction began in 1053 and was completed early in 1060. The bridge, spanning over 800 metres with 47 arches, is made of granites, and each of its beams weighs 20 to 30 tons. The bridge began with the construction of a grillage foundation, and the foundation, as well as the piers, was stabilized by the ubiquitous ostreas. By constructing this bridge, the Chinese wrote a glorious page in the history of bridge building.

Chinese garden architecture is world-famous. The Summer Palace in Beijing and the gardens of Suzhou and Hangzhou enjoy a world reputation.

Silk Weaving   China, historically referred to as Silk Country, is the home of silk. Over 4,000 years ago Chinese people could produce silk cloth from silkworms, indicating that the reeling and weaving of silk had been invented by then. Approximately 2,500 or 3,000 years ago the

On May 18, 1980 China successfully launched a carrier rocket to a predetermined area in the Pacific. (Lower) A helicopter completes its salvaging task and returns to the landing deck of a salvage ship with the instrument capsule.

On September 20, 1981 China launched a group of three space-physics experiment satellites with a single carrier rocket.

China has built her first large high flux atomic reactor. Picture shows scientists at work.

An optical cinetheodolite follows
the path of a flying rocket.

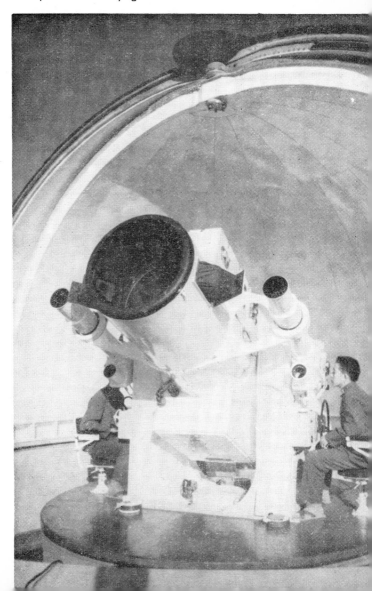

Professor Lu Jiaxi, President of the Chinese Academy of Sciences, studies the theory on the synthesis of a model compound for nitrogen fixation.

The cultivation of hybrid rice which promises increased rice yields is now undertaken on an area of some 4.6 million hectares.

Greenhouse-cultivation of . l improved variety Shandong cotton at Shandong Cotton Institute.

Chinese and foreign scientific researchers on the Qinghai-Tibet Plateau.

Professor Fei Xiaotong, an anthropologist, has received the Huxley medal awarded by the Royal Anthropological Institute of Great Britain. He is the first Chinese scholar so honoured.

Hu Qiaomu, Honorary President of the Chinese Academy of Social Sciences.

念鲁迅诞生一百周年学术讨论会

The centenary of the birth of the Chinese writer Lu Xun is commemorated at this academic conference held in Beijing on September 17, 1981.

The Society of Chinese Historians held a meeting from October 12 to 15, 1981 commemorate the 70th anniversary of the Revolution of 1911 in Wuchang, where th uprising that led to the overthrow of the Qing Government took place. Picture show scholars from various countries attending the reading of papers on the Revolutio figured fabric weaving technology was invented to produce gauze, damask, brocade and other high quality fabrics by the combination of extra warp weave with twill weave. Later, along with the constant improvement of silk weaving, a more satisfactory figured fabric weaving loom came into being.

Through the Silk Road and the sea routes, Chinese silk and silk products were transported to Europe, Asia and Africa where they were well received. Not long afterwards, the technology of silk weaving also spread to foreign countries. For several times between the 5th and 16th century, figured fabric weaving looms were transported to Europe. All this helped the development of textile technology, especially the jacquard technology, in Europe.

Shipbuilding, Maritime Navigation Technology and Invention of Compass During the period between the 2nd century B.C. and the 2nd century A.D., the building of wooden vessels was well developed. The sweep, a high efficient propeller, was created, the stern helm and sail had already been invented, and the vessel itself was put together by the use of iron nails. During this period large ships were built, including the "tower ship" with ten storeys. Around the 3rd century, the sail was improved and made adjustable to wind forces that came from different directions. Another important invention in Chinese shipbuilding was the watertight-compartment construction, which appeared no later than the Tang Dynasty (between the 7th and 9th century), and it was adopted by seagoing ships after the 11th century. The wooden vessels constructed with the watertight compartments and well-caulked with China-made mixture of lime and tung-oil did not leak and provided security against

foundering.   After the 2nd century such vessels plied the
seas between the eastern shore of the Pacific and the
Indian Ocean, even to the eastern coast of Africa. The
seven voyages made by Zheng He, a eunuch in the Ming
Dynasty, was the unprecedented event (1405-33) in the
world's navigation history.   The largest ships Zheng He
employed were about 150 metres long, each equipped
with 12 sails, and the helm pole of such a ship was 11.07
metres long.   An oceangoing voyage of 100 to 200 vessels,
including 40 to 60 ships as large as those described above,
travelling to as far as the eastern coast of Africa, was
indeed a world's record.

The invention of compass and its use in navigation
further sped the development of navigation.   More than
2,000 years ago, the Chinese ground natural magnet in
the shape of a spoon to indicate the north-south direction.
Around the 11th century, the Chinese had discovered two
methods of magnetization.   One was the use of the ter-
restrial magnetic field to magnetize a thin iron piece
shaped like a fish.   As the magnetism gained in this
manner was rather weak, the value of this method was
not high.   Another method was to use natural magnetite
to rub a steel needle which became magnetized in the
process. By making the magnetized needle float on
water, hung or placed in a way as required, it indicated
the north-south direction.   This is the initial form of
compass.   Later on, the compass was fitted with a dial.
Shortly after its invention, the compass was used in navigation and became an important instrument indicating
direction.   In the wake of the compass invention, Chinese
people soon found the magnetic declination which varied
as location changed.

Paper-making and Printing Technology Paper-making in China began in the 2nd century B.C. when people used wornout bast fibres to make rough paper. In the 2nd century writing paper was invented, and the raw materials extended from bast to bark fibres. For several centuries that followed, liana and bamboo fibres were also used for making paper. A handicraftsman, well equipped with necessary tools, could proceed with raw material processing, pulp making, paper making and paper processing, all in one continuous process. China's paper-making technology spread to Korea in the 4th century and Japan in the 10th century, to Arab countries in the 8th century and Europe after the 12th century. Gradually, the whole world learned about it.

Chinese wood-block printing appeared in the period between the 6th and 7th century. It developed rapidly after the 10th century.

The movable-type printing was invented at the beginning of the 11th century. At first books were printed with clay types. At the beginning of the 14th century clay types, and the rotating apparatus for type-setting was invented. After the 14th century tin, copper and lead types were used successively in movable type printing.

Ceramic Manufacture Porcelainware is China's original creation. Porcelain base is mainly made of ceramic clay, with which China produced the first celadon ware over 2,000 years ago but improved its quality to a high level around the 3rd century. Afterwards the porcelain-making technology became more and more sophisticated and porcelainwares of different colours were created, having such colours as white, red, blue, purple and brown. Sometime after the 14th or 15th century, the Chinese mastered the techniques of producing blue and white and

painted porcelainwares. From China, the technology of
making porcelain first spread to Persia in the 11th century. From Persia, it spread to Arab countries. Gradually the whole world learned about it.

Gunpowder The invention of gunpowder began with
the Taoist practice of making elixir pills of immortality
in ancient China. In the 7th century or earlier, elixir pill
makers discovered that a mixture of saltpetre, sulphur
and charcoal could catch fire easily and burn explosively.
Wu Jing Zong Yao (*Collection of the Most Important
Military Techniques*), compiled in 1044 during the Northern Song Dynasty, mentions three kinds of gunpowder
(poisonous gunpowder, incendiary gunpowder and explo-
sive gunpowder) and records their composition in detail.

Gunpowder was used in fighting wars as early as the
10th century. In the 12th century at the latest, the tube-
shaped firearms were invented. From the 13th century
onwards, the knowledge of making gunpowder spread to
Arab countries wherefrom it went to Europe.

Like gunpowder, the practice of making elixir pills,
an initial form of chemical research prior to the formation of modern chemistry, emerged in China around the
beginning of the Christian era, and, like gunpowder, also
went to Europe via Arab countries. In addition, China
learned to burn coal, petroleum and natural gas earlier
than any other country in the world.

2. GENERAL SURVEY OF SCIENTIFIC AND TECHNOLOGICAL DEVELOPMENT IN NEW CHINA

As has been stated previously, China made significant
contribution to the world in science and technology in

the past, but in modern times the situation is somewhat different. In about 100 years before the founding of New China in 1949, natural science stagnated because the imperialist invasion from abroad and the corruption of reactionary rulers at home combined to fetter the productive forces in the nation. On the eve of Liberation, there were only just over 600 persons engaged in scientific research and the research areas were anything but complete. The research areas of newly developed science that were closely linked with production were practically non-existent. In old China, many scholars went abroad to study sciences in the hope of using the acquired knowledge to help China after they returned: to initiate new scientific enterprises, popularize modern scientific knowledge and train scientific personnel. Their contribution should not be minimized.

Soon after the founding of New China, in order to develop China's science and technology, the state took over and reorganized the research organs of the Central Research Academy and the Beiping Research Academy previously under the Kuomintang government and established the Chinese Academy of Sciences in November 1949. At its founding, the Chinese Academy of Sciences had only about 20 research institutes with some 200 research persons. Later, institutes for studying newly developed sciences were established successively, new instruments and equipment were added, a large number of new people were recruited, and more scientific researchers were trained quickly. In 1955, the Chinese Academy of Sciences had under its jurisdiction 44 research institutes with 2,485 researchers. At the same time various government departments set up their own research organizations successively, and industrial departments established institutes or laboratories in important industrial enterprises. Having undergone adjustment and reform, Chinese colleges and universities paid great attention to scientific research which was elevated to a new status. In the first half of 1955, there were in China 194 colleges and universities with a teaching staff of 42,000, who constituted another important scientific research force. The co-operation and co-ordination among various scientific groups, each sharing a part of the research work, formed a nationwide research system in China.

In order to meet the need arising from the rapid development of socialist construction, the Central Committee of the Communist Party of China held a conference in January 1956 to solve all problems concerning intellectuals. In the conference, the function of intellectuals in the context of socialist construction was affirmed and correct policies towards them were approved. The conference, emphasizing the role of science and technology in modernization, issued the call, "March Towards Science". In June of the same year, in order to promote socialist science and cultural activities, the Party Central Committee proclaimed the general policy of "Let a hundred flowers blossom and a hundred schools of thought contend". All this helped accelerate the development of Chinese science and technology. Shortly afterwards, the State Council set up the Science and Technology Planning Commission, mobilizing more than 600 scientific and technological experts to work out the first long-term programme, the 12-year Programme for Scientific and Technological Development (1956-67). This programme listed atomic energy, jet technology, computer, semiconductor, electronics, and automatic control as main areas for development. It paid close attention to the applied science while at the same time emphasizing the importance of basic science and theoretical research. Applied science was thus combined with basic science, and the present need with the long-term arrangement. The main projects in this programme were completed in 1962, five years ahead of schedule. In 1956, the State Council set up the State Technology Commission. In 1958, it was merged with the Science and Technology Planning Commission to form the State Science and Technology Commission which, from then on, had jurisdiction over science and technology regulations, plans and policies. In 1962, the new commission formulated a 10-year programme (1963-72), which was not carried out entirely because of the interference by the "cultural revolution". The implementation of the plans for scientific and technological development played an important role in the growth of Chinese scientific and technological personnel, the raising of China's scientific level, the filling of gaps in science and technology and the solution of scientific and technological problems arising from production and construction.

For 17 years from 1949 to 1965, Chinese science and technology developed rapidly. In 1965, China had over 120,000 researchers in more than 1,700 scientific research institutions, 106 of which, with about 22,000 researchers, belonged to the Chinese Academy of Sciences. The Chinese Science and Technology Association, established in 1958, had 53 academic societies, which altogether published 92 academic periodicals. Within 17 years, China developed a series of new disciplines, such as atomic energy, electronics, semiconductor, automatic control, computer, jet technology, laser, and infrared ray. It enriched and raised the academic level of such basic sciences as mathematics, physics, chemistry, astronomy, geography, geology and biology. As a result of more than a decade's work, the gap in science and technology between China and advanced countries in the world was narrowed.

However, during the "cultural revolution" from 1966 to 1976, development of science and technology in China suffered enormously because of the serious interference and sabotage by the two reactionary cliques of Lin Biao and Jiang Qing. In 1973 only 53 research institutes with just over 13,000 researchers remained in the Chinese Academy of Sciences. The Chinese Science and Technology Association and its subsidiary organizations stopped activities altogether, including publication of their respective journals. Meanwhile, schools ceased receiving new students for a period of four years; as a result, China lost more than one million college students and nearly all research work in colleges and universities stopped. All this affected adversely China's scientific and technological development: the number of Chinese scientific and technological personnel reduced, their academic level lowered, their average age became older and older. The younger researchers were not qualified enough to take up the jobs left over by those who were about to retire. With the exception of a few fields, scientific research virtually stopped during this period. The gap in science and technology between China and the more advanced countries widened.

Since 1976, China has entered upon a new historical period. The central task in this period is to modernize agriculture, industry, national defence, and science and technology. The scientific and technological modernization is crucial to the realization of the other three modernizations. To speed up China's scientific and technological development, the state, in March 1978, held a national conference on science that was attended by 6,000 representatives. The conference formulated an eight-year plan (1978-85) which singled out eight areas of research as the key that would affect the overall situation. The eight areas are agriculture, energy, materials, electronic computer, laser, space, high-energy physics and genetic engineering. Meanwhile, a large number of scientific research organizations, experimental and survey centres and scientific and technological services were restored or established, giving impetus to further development in this field. In 1980, the scientific and technological community, basing upon the experience gained after the founding of new China, cleared up the "Leftist" ideology and its influences and formulated a new policy for scientific development. The basic content of this policy is as follows: First, the development of science and technology should be well co-ordinated with the development of economy and society, and the development of economy shuold be the first priority. Second, at the present moment stress should be placed on the research work involving productive technology, namely, the selection of needed techniques and the formation of a rational technological structure. Third, in factories and mines the application of new techniques and their promotion should be strengthened. Fourth, research on basic sciences should be assured of stable development. Lastly, on the principle of self-reliance and independence, scientific and technological attainments among the advanced countries in the world must be earnestly studied and made useful to China. In short, the focus of this policy is that the development of the national economy must rely on science and technology which in turn must serve the

development of the national economy. Now all scientific and technological workers in China are doing their
utmost to implement this general policy.

For more than 30 years, China has established a
scientific and technological research system of five component parts of over 4,000 research organizations. The
first part is the Chinese Academy of Sciences with 177
research organizations. It is a comprehensive, nationwide research centre of natural science, which stresses
research on basic science, the level of which it aims to
raise. Ultimately, it serves the nation's economic construction and defence. The second part is the scientific
research and planning organizations that function under
the various departments of the State Council or those on
the local level, such as the Academy of Agricultural
Sciences, Academy of Forestry Science, Academy of Medical Sciences, Research Institute of Traditional Chinese
Medicine, Academy of Geological Science, Research Institute of the Ministry of Railways, General Academy
of Iron and Steel, Scientific Research Institute of
Petroleum Prospecting and Exploitation, and Research
Institute of Textile Science. Their main task is to engage
in applied research closely related to the needs and
characteristics of the departments or regions. The third
part is the research organizations that operate under institutions of higher learning. Many colleges and universities have established research institutes or offices of
various kinds. For instance, the well-known Beijing
University has the research institutes of mathematics,
solid state physics, theoretical physics, heavy ion physics,
physical   chemistry,   molecular   biology,   computer
science, remote-sensing technology, etc. China's institutions of higher learning, especially the key col-

leges and universities, are centres of not only teaching and study but also scientific research. Their research areas consist of basic science, applied science and important subjects involving national economy and defence. The fourth part is the research organizations run by factories and mines. Their main concern is projects connected with the development of production and construction. Many large factories and mines have set up their own research institutes or offices. The fifth part is the research organizations of the national defence, which concern themselves mainly with the newly developed techniques needed for national defence. These component parts of the nationwide scientific research system work independently, but they also cooperate with one another.

In 1980 China had 5.296 million scientific and technological workers associated mainly with the research organizations described above. The general situation is summarized in the table on p. 144.

Apart from government research organizations, there are 115 nationwide academic societies under the jurisdiction of the Chinese Science and Technology Association. They have more than 2,000 branches in large and medium-sized cities, forming an extensive network of scientific and technological exchange. In 1980 the Chinese Science and Technology Association held the second national congress, and its affiliated academic societies organized many activities, which were helpful in promoting Chinese scientific and technological development. In the same year, academic societies held more than 700 large-scale academic conferences to discuss important scientific and technological problems relating to the four modernizations. The attendance totalled more than 80,000

Number / Item \ Year	1952	1960	1978	1979	1980
Scientific Research Workers	8,000	90,500	310,300	316,800	322,900
Engineers and Technicians	164,000	820,700	1,571,200	1,666,900	1,862,200
Agricultural Technicians	15,000	167,300	294,200	324,600	311,100
Medical Workers	126,400	517,800	1,275,600	1,396,400	1,530,200
Teachers	111,600	372,600	893,800	1,000,700	1,249,900
Scientific and Technological Editors and Translators.					19,700
Total	425,000	1,968,900	4,345,100	4,705,400	5,296,000

persons who presented more than 100,000 papers. The popularization of science is an important task for all the academic societies. Under the unified arrangement of the association, they organized a wide range of activities such as publicity given to the four modernizations, circular reports, lectures on technology, short-term training courses, and various exhibits.

Because of a weak foundation, shortcomings in our work, errors in our policies and especially the interference and sabotage by the counter-revolutionary cliques of Lin Biao and Jiang Qing, China's science and technology still lag behind the advanced level of the world. However, as China has built up its own scientific and technological research system and trained a number of scientific and technological workers of considerable knowledge and skill, it has accomplished a great deal in important scientific and technological fields. In 1980 China obtained good results in 2,600 scientific and technological items, 1,576 items of which, including 329 important ones, originated in the Chinese Academy of Sciences. The successful experiment with atomic and hydrogen bombs and also with missiles, the launching of satellites and their safe return, and the synthesis of bovine insulin — all this symbolizes the new, higher standards of Chinese science and technology. On September 20, 1981, China, for the first time in its history, launched successfully a group of three spacephysics experiment satellites with a single carrier rocket. Now China has acquired the capability to solve many important and complex problems in science and technology that exist in national construction, including some of the most difficult ones that require the most advanced techniques.

3. MAJOR SCIENTIFIC AND TECHNOLOGICAL ACHIEVEMENTS IN MORE THAN THREE DECADES

Agricultural Science In agricultural science Chinese scientific workers have made important contributions in production as well as scientific research. For instance, they improved and learned to utilize alkaline and red loam, and other low-quality soil. They successfully researched the selection and breeding of good strains of rice, wheat, cotton and other main crops, and they studied the laws that governed high yielding. They successfully cultivated hybrid rice and all-octoploid black wheat, and they did research on the occurrence of locust plague and learned about its prevention and elimination. They worked on a comprehensive method of preventing and eliminating wheat rust and midge, and they strove to find out about the laws that governed the migration of armyworms. They successfully made rinderpest and hog cholera lapinized vaccines. They diagnosed horse anaemia and learned about its immunity, and they artificially bred black, grass, silver and big-head carps. Artificially too, they raised kelp and laver seedlings, and they promoted utilization of methane in rural areas.

Technological Science As the national economy and defence requirements continued to develop, major efforts have been made in the areas of mining, ore dressing, metallurgy, machinery, electric engineering, civil construction, water conservancy and electric power projects, chemical engineering, thermal physics, materials science, applied dynamics and optics. Meanwhile, gaps have been filled in the following fields: electronics, computer, semiconductor, automatic control, laser and infrared ray, space technology, corrosion and its prevention. The completion of the Changjiang River Bridge at Nanjing, the Chengdu-Kunming Railway and many other big projects, the comprehensive utilization of vanadium and titanium-bearing magnetite and rare-earth ore, the successful development of alloy steel and nonferrous alloy by using Chinese resources, the successful manufacture of the 12,000-ton hydraulic press, the huge generator with an inner water-cooled stator and rotor, the large-capacity electronic computer, the large-scale integrated circuit, the high resolving power electron microscope and several large precise surveying optic installations, and the achievements made in the three-dimensional flow theory for calculating turbine machinery, the multibumping theory for calculating the strength of machine parts, and the metal-break theory — all this indicates that Chinese research in technological science has been crowned with great success.

Medical Science (See *Sports and Public Health* of the "China Handbook Series".)

Mathematics China has a long history of achievements in mathematics. Since the founding of New China pure mathematics has continued to make progress. Particularly, research results on the additive theory of numbers, the sieve method and Goldbach's Conjecture have been universally recognized and accepted. Important contributions have been made in topology, and function theory, group theory, differential equation and functional analysis have yielded equally good results. In applied mathematics China has independently developed the finite element method and its theory. Research work on the information theory and the mathematic theory of computer science has begun. The mathematical statistics and mathematical logic, the probability theory and the operational research have made rapid progress while the spread of optimum seeking method has achieved success. Excellent results have also been made in cybernetics.

Astronomy For a period of over 30 years China has established the Beijing, Shaanxi, Yunnan and other astronomical observatories, astronomical instrument plants, and a network of observatory stations for artificial satellites. The time and pole-move services, the solar activity forecast, and the making of astronomical calendar have all begun in order to meet the urgent need of national economy and defence. In 1963, the precision of China's service for universal time was ranked among the world's most advanced. The research on the physics of the sun, fixed stars and galaxies, and the universe as a whole was extended gradually. In recent years theoretical research has yielded effective results.

Physics The theoretical research and experimental work on solid state physics, nuclear physics, high-energy physics, optics, acoustics and other sciences have had encouraging results. Many people engaged in physics research have directly or indirectly participated in the work of manufacturing atom and hydrogen bombs, missiles and artificial earth satellites, while even more people have made contributions to the development of China's machinery, electronics, metallurgical and instrument industries, to the establishment of China's atomic energy, semiconductor and other newly developed industries, and to the various other aspects of national economy and defence. In 1959, Chinese physicists found the negative hyperon against sigma and in 1966 advanced the straton model theory. In 1972 the Yunnan Cosmic Ray Station located a heavy particle which may be ten times heavier than proton. In recent years the Samuel Ting group, joined by Chinese high-energy physicists, successfully proved the existence of gluon. Meanwhile, the Wei Mo group, together with Chinese high-energy physicists, made the important progress on the unity of weak interaction and electromagnetic interaction through the experiment of mu-neutrino and electron elastic scattering. These are only two examples showing how China has successfully co-operated with the international scientific and technological community.

Chemistry During the past 30 years or more, new branches of chemistry have been set up, such as theoretical chemistry, structural chemistry, high polymer chemistry, high polymer physics, organic synthesis, catalytical and element organic chemistry, and colloidal chemistry. Research achievement with China's distinguishing features has also come about. The synthesis of insulin and the crystalline structural analysis have reached the international advanced level. The production of aviation kerosene, major catalytic agents used in petroleum and chemical industries, synthetic butadiene rubber and rare-earth isoamyl rubber, synthetic mica, silicon and fluorine materials has been instrumental in China's industrial modernization, especially in the development of synthetic rubber and petrochemical industries. China's industry also provides the atom and hydrogen bombs and man-made satellites with new crucial materials such as solid missile propellant, fuel cells and solar energy cells, temperature-controlled coating materials, and synthetic rubbers, plastics and lubricants that are resistant to high temperature, low temperature and

strong oxidizer. It has made its contribution to national
defence. In recent years outstanding progress has been
made in the synthetic chemistry, structural chemistry,
quantum chemistry, catalytic and rare-earth chemistry
and analytical separation. For instance, the linear law
of organic homologue and the active centre model of
nitrogen-fixing ferment are advanced further, the rare-
earth extraction and separation and mini-chromatogram
technology are successfully adopted and the efficient
deoxidation catalyst and the high-temperature-resistant
burn-through materials are produced.

Biology Before Liberation the research work in
biology remained descriptive; since then systematic and
experimental research has made rapid progress and some
important branches of biological science have been set up
or strengthened. Meanwhile, much research has been
done to meet the needs of industry, agriculture, medicine
and national defence. The utilization of hybrid vigour,
the hereditary breeding, the prevention and control of
locust plague, the biological prevention and elimination of
plant diseases and insect pests, the stimulation of oestrus
for cultured fish, the cultivation of laver and kelp, the
use of new-type microbial ferment and the fracture fluid
of cyamopsis tetragonolobus and sesbania of mucilage, the
analgesia tenet of acupuncture anaesthesia, the utilization of Chinese medicinal herbs, the investigation and
compiling of illustrated books on biota and soil and the
domestication of castor silkworm and other animals and
plants — all this has contributed to the cause of China's
socialist construction.

Earth Science The geophysics, atmospheric physics,
marine physics, marine chemistry, marine geology, geo-
chemistry, engineering geology and hydrogeology have all

been established or strengthened, and a comprehensive investigation of China's seas and littoral zone has been carried out. In recent years China has made great progress in the investigation of continental shelves along the west Pacific, the South China and East China seas, including investigation of ocean contamination. The nationwide meteorological observatory network has been established, and new progress is being made in the basic research of dynamic meteorology. The geotectological theory has provided guidance for the development of geology and the forecast of mineralization. The nationwide network of seismic stations has been set up; the earthquakes of Haicheng, Longling and Songpan were successfully forecast. Much survey has been done along the Heilongjiang River and the middle reaches of the Huanghe River and in Xinjiang, Qinghai, Inner Mongolia, Ningxia and other regions. Based upon these surveys, an atlas of China's natural environment has been compiled. The control of glacier and desert, frozen soil in Qinghai and Tibet, and mud-rock flow in southwest China — all this has been studied systematically. In recent years an on-the-spot investigation in the Mount Qomolangma region and research on the rise of the Qinghai-Tibet Plateau — its process, cause and impact upon natural environment — has made considerable progress.

Some Information About the Man-Made Satellites Launched by China On September 20, 1981, for the first time in its history, China successfully launched a group of three spacephysics experiment satellites with a single carrier rocket. These are the ninth, tenth and eleventh satellites launched by China.

Prior to this, China had successfully launched eight man-made earth satellites:

The first man-made satellite was sent aloft on April 24, 1970.

Then on March 3, 1971, a scientific experiment satellite was sent into space.

After that, China launched six more satellites respectively on July 27, November 26 and December 16, 1975, on August 30 and December 7, 1976, and on January 26, 1978.

According to data that have been made public, the first man-made earth satellite, launched on April 24, 1970, weighed 173 kilogrammes. It made one complete revolution round the earth in 114 minutes along a trajectory the perigee of which was 439 kilometres and the apogee 2,384 kilometres. The angle of its orbit to the equator plane was 68.5°. The scientific experiment satellite launched on March 3, 1971, weighed 221 kilogrammes. It made one complete revolution round the earth in 106 minutes along a trajectory the perigee of which was 266 kilometres and the apogee 1,826 kilometres. The angle of its orbit to the equator plane was 69.9°. The satellite launched on July 27, 1975, made one complete revolution round the earth in 91 minutes along a trajectory the perigee of which was 186 kilometres and the apogee 464 kilometres. The angle of its orbit to the equator plane was 69°. The one launched on November 26, 1975, made one complete revolution round the earth in 91 minutes along a trajectory the perigee of which was 173 kilometres and the apogee 483 kilometres. The angle of its orbit to the equator plane was 63°.

The satellite sent on November 26, 1975, returned to earth according to plan and the one launched on December 7, 1976, also returned right on schedule. The one launched on January 26, 1978, having completed the planned experiments, also returned to earth uneventfully.

4. ORGANIZATIONS FOR SCIENTIFIC RESEARCH

State Science and Technology Commission The State Science and Technology Commission is a government organ under the jurisdiction of the State Council; it is responsible for scientific and technological work. Its basic tasks are to formulate policies in scientific and technological matters; to decide on the main scientific and technological projects to be undertaken and make plans for scientific and technological development; to guide and assist various departments and regions in their scientific and technological activities; to distribute funds for the manufacture of new products, on an experimental basis, as well as important scientific research items and major intermediate experiments; to supervise scientists and technicians; to appraise, register and promote scientific and technological results; to supply scientific and technological work with required materials; to co-ordinate the scientific and technological exchange and co-operation with foreign countries.

Every province, municipality and autonomous region have their own scientific and technological commissions as administrative organs responsible for the guidance of scientific and technological work.

In order to promote scientific research, the State Science and Technology Commission has established a China Institute of Scientific and Technological Information responsible for collecting, collating, preserving, indexing, translating and publishing scientific and

technological information both at home and abroad.
Similar information centres have also been established in
various departments and regions. Besides the State
Science and Technology Commission, there are various
kinds of service corporations, such as the China Scientific
Equipment Corporation, China Publications Import and
Export Corporation, China Scientific and Technological
Document Translation Corporation, and China Scientific
and Technological Audio-Video Service Corporation.

Chinese Academy of Sciences   The Chinese Academy
of Sciences is the supreme academic organ and the comprehensive research centre of natural sciences. Its principal objectives are to conduct basic scientific and theoretical research and to conduct research into and develop
new branches of science and technology so as to serve the
national economy and defence. Its tasks are to run its
research organization well, and through the research
work in basic, technological and newly developed sciences
to gain scientific and technological results with academic
value as well as practical importance. It trains scientists
and technicians who love their socialist motherland and
are full of creative talents and indomitable spirit. It seeks
to ascertain the development orientation of each scientific
branch, to strengthen the connection, interchange and co-
operation of its research organizations with the research
organizations in institutions of higher learning, various
governmental departments or localities, and to play an
important role in the popularization of science. It acts
as an advisory body to the Party and the government on
general and specific policies and on important scientific
and technological problems that arise in the course of
socialist construction or modernization. As an academic
organ of the state, it participates in important academic

events throughout the world and engage in academic exchange and co-operation with other countries.

According to the Provisional Constitution of the Chinese Academy of Sciences, the Academicians' General Meeting is the supreme policy-making body of the Chinese Academy of Sciences. Its functions are to discuss and decide on the development orientation of the academy; to examine or revise research plans; to make decision on important questions; and to elect a presidium. Academicians are chosen from among outstanding scientists of the country and elected at the academicians' conference of different academic departments. Today there are altogether 400 academicians. Normally the Academicians' General Meeting is held once every two years. The latest one, the Fourth Academicians' General Meeting, was held from the 11th to the 20th of May 1981. The presidium is the policy-making organ when the Academicians' General Meeting is not in session. A presidium member is elected for a term of four years and can be re-elected once, while the president and vice-presidents, chosen and elected among members of the presidium for a term of two years, can also be re-elected once. The president and vice-presidents assume leadership in daily administrative matters.

The establishment of academic departments within the Chinese Academy of Sciences is an important measure of relying on scientists to manage scientific development work. An academic department is mainly an administrative organ, but it also assumes to some extent the management of scientific research in its affiliated research institutes. The department director, vice-directors and the standing committee take care of the daily administrative matters. Today there are altogether five academic departments: physics and mathematics, chemistry, biology, earth science, and technological science.

The basic organs within the Chinese Academy of Sciences are its research institutes, the directors and vice-directors of which are responsible for both scientific research and administrative work. At present, there are 177 scientific research institutes.

Responsible Members of the Chinese Academy of Sciences Members of the Presidium (in the order of the number of strokes in their surnames): Yu Guangyuan, Wang Daheng, Wang Ganchang, Ye Duzheng, Feng Depei, Lu Jiaxi, Hua Luogeng, Song Ping, Yan Dongsheng, Yan Jici, Li Chang, Li Xun, Wu Zhonghua, Wu Zhengyi, Wu Heng, Yu Wen, Zhou Peiyuan, Zhang Wenyou, Zhang Guangdou, Hu Keshi, Hou Xianglin, Qin Lisheng, Qian Sanqiang, Qian Xuesen, Tu Guangchi, Gao Yi, Tang Aoqing, Huang Jiasi, Xie Xide

Executive Chairmen: Yan Jici, Li Chang, Wu Zhonghua

President: Lu Jiaxi

Vice-Presidents: Qian Sanqiang, Hu Keshi, Feng Depei, Li Xun, Yan Dongsheng, Ye Duzheng

Academic Department Directors:

Qian Sanqiang, Director of the Department of Physics and Mathematics

Yan Dongsheng, Director of the Department of Chemistry

Feng Depei, Director of the Department of Biology

Tu Guangchi, Director of the Department of Earth Science

Li Xun, Director of the Department of Technological Science

Research Organs Under the Direct Jurisdiction of the Chinese Academy of Sciences

Name	Location
Institute of Mathematics	Beijing
Institute of Applied Mathematics	"
Institute of Systematic Science	"
Institute of Mechanics	"
Institute of Physics	"
Institute of High Energy Physics	"
High Energy Physics Experiment Centre	"
Institute of Theoretical Physics	"
Institute of Acoustics	"
Institute of Chemistry	"
Institute of Chemical Metallurgy	"
Institute of Photosensitive Chemistry	"
Institute of Environmental Chemistry	"
Beijing Astronomical Observatory	"
Institute of Geography	"
Institute of Atmospheric Physics	"
Institute of Vertebrate Paleontology and Paleoanthropology	"
Commission of Comprehensive Survey of Natural Resources	"

Institute of Geophysics	Beijing
Institute of Geology	"
Institute of Biological Physics	"
Institute of Microbiology	"
Institute of Genetics	"
Institute of Psychology	"
Institute of Zoology	"
Institute of Growth Biology	"
Institute of Botany	"
Institute of Computing Technology	"
Computer Centre	"
Institute of Thermal Physics	"
Institute of Electronics	"
Institute of Semiconductor	"
Institute of Automation	"
Institute of Electrical Engineering	"
Space Science and Technology Centre	"
Institute of Remote-Sensing Application	"
Institute of the History of Natural Science	"
Shanghai Institute of Atomic and Nuclear Energy	Shanghai
Shanghai Institute of Organic Chemistry	"
Shanghai Institute of Silicate Chemistry and Engineering	"

Shanghai Astronomical Observatory	Shanghai
Shanghai Institute of Biochemistry	"
Shanghai Institute of Medicine	"
Shanghai Institute of Cell Biology	"
Shanghai Institute of Physiology	"
Shanghai Institute of Plant Physiology	"
Shanghai Institute of Insectology	"
Shanghai Institute of Metallurgy	"
Shanghai Institute of Optical and Precision Machinery	"
Shanghai Institute of Technical Physics	"
Shanghai Institute of Brain Study	"
Zijinshan Astronomical Observatory	Nanjing
Nanjing Institute of Geology and Paleontology	"
Nanjing Institute of Pedology	"
Nanjing Institute of Geography	"
Anhui Institute of Optical and Precision Machinery	Hefei
Hefei Institute of Plasma Physics	"
Hefei Institute of Intelligent Machinery	"
Fujian Institute of Substance Structure	Fuzhou
Institute of Oceanography	Qingdao
Shijiazhuang Institute of Agricultural Modernization	Shijiazhuang

Shanxi Institute of Coal Chemistry	Taiyuan
Chengdu Institute of Organic Chemistry	Chengdu
Chengdu Institute of Geography	"
Chengdu Institute of Biology	"
Institute of Optical and Electrical Technics	Dayi, Sichuan
Chengdu Institute of Computer Application	Chengdu
Yunnan Astronomical Observatory	Kunming
Kunming Institute of Zoology	"
Kunming Institute of Botany	"
Yunnan Institute of Tropical Botany	Mengla, Yunnan
Guiyang Institute of Geochemistry	Guiyang
Guangzhou Institute of Chemistry	Guangzhou
Institute of South China Sea Oceanography	"
Institute of South China Plants	"
Guangzhou Institute of Electronic Technology	"
Guangzhou Institute of Energy Resources	"
Guangzhou Institute of New Technology in Geology (in preparation)	"
Guangzhou Observatory for Man-Made Satellites	"
Changchun Institute of Physics	Changchun
Changchun Institute of Applied Chemistry	"

Changchun Institute of Geography	Changchun
Changchun Institute of Optical and Precision Machinery	"
Changchun Observatory for Man-Made Satellites	"
Harbin Institute of Precision Instruments	Harbin
Heilongjiang Institute of Agricultural Modernization	"
Institute of Metals	Shenyang
Institute of Forestry and Pedology	"
Shenyang Institute of Computing Technology	"
Shenyang Institute of Automation	"
Dalian Institute of Chemical Physics	Dalian
Wuhan Institute of Rock Mechanics	Wuchang
Wuhan Institute of Physics	"
Wuhan Institute of Mathematics and Physics	"
Wuhan Institute of Viruses	"
Wuhan Institute of Botany	"
Wuhan Institute of Hydrobiology	"
Institute of Survey and Geophysics	"
Changsha Institute of Earth Structure	Changsha
Changsha Institute of Agricultural Modernization	"
Institute of Modern Physics	Lanzhou
Lanzhou Institute of Chemical Physics	"

Lanzhou Institute of Glacier and Frozen Soil	Lanzhou
Lanzhou Institute of Desert	"
Lanzhou Institute of Highland Atmospheric Physics	"
Lanzhou Institute of Geology	"
Xinjiang Institute of Physics	Urumqi
Xinjiang Institute of Chemistry	"
Xinjiang Institute of Biology, Soil and Desert	"
Xinjiang Institute of Geography	"
Urumqi Observatory for Man-Made Satellites	"
Xian Institute of Optical and Precision Machinery	Xian
Shaanxi Astronomical Observatory	Pucheng, Shaanxi
Institute of Space Physics	Xian
Northwest China Institute of Water and Soil Conservation	Wugong, Shaanxi
Qinghai Institute of Salt Lakes	Xining
Northwest China Institute of Highland Biology	"

Chinese Science and Technology Association The Chinese Science and Technology Association has as its members scientists' and technicians' organizations, including special academic bodies, research societies and associations, whose members are mostly scientists and scientific and technological workers. Its aim is to advance the development and success of science and technology in China, popularize scientific knowledge among the

Chinese people, and raise their standard of knowledge about science and technology. In short, the aim is to raise the scientific and cultural level of the entire nation and make China a modern socialist country as early as possible. Its tasks are to organize and support its members in academic interchanges and to compile and issue academic publications. It makes use of various forms to popularize the knowledge of science and technology among cadres and the masses, provides scientific and technological education for youngsters, and actively assists relevant departments to make technological exchanges and general scientific experiments. It acts as an adviser to government departments, enterprises and institutions and undertakes the tasks that have been entrusted to it by them. It encourages its members and other scientists and technicians to make suggestions on scientific and technological development. It promotes activities of popularizing science and technology and also activities that serve scientists and technicians. It transmits, in time, opinions, suggestions and demands of scientists and technicians to proper authorities. It participates actively in international exchange programmes and promotes the friendly ties between Chinese scientists and technicians and their counterparts in foreign countries.

The Chinese Science and Technology Association has its branch organizations at the levels of provinces, centrally administered municipalities, and autonomous regions, provincially administered municipalities, regions, counties, factories, mines, and people's communes. Today it has 115 special academic bodies, research societies, and associations. It has also more than 2,000 branches in large and medium-sized cities and published over 130 periodicals.

Responsible Members of the Chinese Science and Technology Association

Chairman: Zhou Peiyuan

Vice-Chairmen: Pei Lisheng, Qian Xuesen, Huang Jiasi, Liu Shuzhou, Yan Jici, Mao Yisheng, Hua Luogeng, Zhang Wei, Lin Lanying, Yang Xiandong, Yang Shixian, Qian Sanqiang, Jin Shanbao, Wang Ganchang, Wang Shuntong

5. INTERNATIONAL CO-OPERATION AND EXCHANGE IN SCIENCE AND TECHNOLOGY

The Chinese government attaches importance to the strengthening of scientific and technical co-operation with foreign countries and holds that the unequal relations and unreasonable restrictions in international scientific and technological contacts should be abolished forthwith.

For a long time China has co-operated with the developing countries in many fields, such as agriculture, forestry, water conservancy, electricity, light industry, food processing, textiles, mechanical engineering, metallurgy, chemical engineering, architecture, and transport and communication.

China has signed agreements on scientific and technological co-operation and agreements on economic and industrial co-operation with the following countries: Korea, Argentina, Pakistan, Bangladesh, Thailand, the Philippines, Libya, Zambia, Mexico, Chile, Romania, Yugoslavia, Hungary, Poland, Czechoslovakia, Bulgaria, the German Democratic Republic, France, Britain, the Federal Republic of Germany, Italy, Sweden, the United States, Greece, Denmark, Finland, Belgium, Luxemburg, Australia, Japan, Norway, etc. These agreements include extensive co-operation in basic research, agricultural and industrial production and newly developed technologies.

The co-operation with the scientific and technological organizations of the United Nations is an important part of China's international co-operation. These organizations include UNESCO, UNIDO, UNCTAD, UNFAO, WHO and ILO. Since 1979 China has actively supported and participated in the United Nations Conference on Science and Technology for Development and the Centre for Science and Technology for Development.

Since 1980 the exchange of visits between Chinese scientists and their foreign counterparts has increased substantially. The Chinese Science and Technology Association alone has hosted 21 invited delegations with a total of 132 members, in addition to more than 300 scientists who came to China as tourists. It also sent 22 delegations of 130 members to foreign countries. In 1980, 13 Chinese academic societies joined relevant international organizations; thus the number of Chinese academic societies that have joined international academic organizations now numbers 61. Ministries and commissions that function under the central authorities, institutions of higher learning and the Chinese Academy of Sciences also make their own arrangements in scientific exchanges on a large scale. In 1980, for instance, the Chinese Academy of Sciences sent delegations and observation groups to foreign countries and participants to international conferences, totalling 950 persons. Meanwhile, it played host to 1,760 foreign scientists, 50 per cent more than in 1979. In the same year, the academy sent to foreign countries 674 postgraduates, students for advanced study and visiting scholars.

In 1980, 19 international academic conferences were held in China, such as the Conference on Particle Physics, International Mine Planning and Development Symposium, Symposium on the Qinghai-Tibet Plateau, International Conference on Lasers, and Symposium on Paddy Soil.

Many famous Chinese scientists have received honourable titles from foreign countries. For instance, Guo Kexin was appointed a foreign academician by the Royal Swedish Academy of Engineering Sciences. Zhang Xiangtong received a Threshold Award. Some of China's academic and scientific organizations also give honourable titles to foreign scholars or invite them to hold posts. For instance, the Institute of System Science of the Chinese Academy of Sciences invited Academician Shiing Shen Chern and Professor Lax of the National Academy of Sciences in the United States as honourable research professors and members of the science commission. Meanwhile, the Institute of Mathematics invited Shing Tung Yau, a tenured professor at Princeton University, and Shiu Yuen Cheng, an associate professor at University of California in Los Angeles, as members of the science commission.

Besides, Chinese scientists have conducted research jointly with foreign scholars. For instance, China and France have jointly studied the geological structure of the Himalayas. China has published the Chinese edition of the American magazine *Scientific American* and also the American book entitled *Encyclopedia of Science and Technology*.

Chapter Three

SOCIAL SCIENCES

1. SOCIAL SCIENCES IN CHINA

From Ancient Academic Thinking to Modern Social Sciences In China, the term "social sciences" covers humanities as well as social sciences. "Social sciences" in its present-day meaning has a history of only about a century. However, the nascent thought of philosophy, aesthetics, political science, the science of law and economics dates back to the Spring and Autumn Period (770-476 B.C.), or as early as the period between the Shang and Zhou dynasties (about 11 centuries B.C.). Books appeared even earlier, for during the Shang period (about 16-11 centuries B.C.) there were official historians in charge of books. Towards the end of the Spring and Autumn Period and in the Warring States Period (475-221 B.C.), there were drastic social upheavals along with the transition from a slave to a feudal society. The official monopoly of academic pursuit was broken, and individuals began to preach and write about their own doctrines. This marked the beginning of "a hundred schools of thought contending". Confucius, Lao Zi, Zuo Qiuming, Sun Wu, Mo Di, Shang Yang, Mencius, Hui Shi, Zhuang Zi, Gongsun Long, Xun Qing and Han Fei were the most accomplished among a large number of philosophers, ethicians, logicians, aestheticians, political scientists, jurists, military scientists, economists, historians and educationalists. An abundance of academic writings emerged, such as *Lun Yi* (*The Analects of Confucius*) which is a Confucian classic that records the words and deeds of Confucius and his pupils; the Taoist School classics *Lao Zi* (*The Book of Lao Zi*) and *Zhuang Zi* (*The Book of Zhuang Zi*); the Legalist School classics *Shang Jun Shu* (*The Book of Shang Yang*) and *Han Fei Zi* (*The Book of Han Fei*); *Zhou Yi* (*The Book of Changes*) which contains rudimentary materialism and dialectics; *Shang Shu* (*The Book of Historical Documents*) which is an anthology of ancient state archives; and the earliest Chinese historical writings *Chun Qiu* (*Spring and Autumn Annals*), *Zuo Zhuan* (*Zuo Qiuming's Chronicles*), *Guo Yu* (*Anecdotes of the States*) and *Zhan Guo Ce* (*Sayings of the Warring States*). The period during which these books flourished was the dawn of social sciences in China.

In the Qin (221-207 B.C.) and Han (206 B.C.-A.D. 220) dynasties, China was already a unified, multi-nationality state under a centralized feudal autocracy. Emperor Wu Di (r. 141-87 B.C.) of the Han Dynasty adopted and promoted a policy of "banning the hundred schools and elevating alone Confucian teachings". Henceforth, Confucianism became the orthodox ideology. Nevertheless, different schools of thought continued to develop, changing their contents with the change of time. Over the long feudal era, thinkers and scholars appeared in succession, and they exercised great influence on their successors.

Drawing lessons from the short-lived Qin Dynasty, the rulers of the early Western Han Dynasty (206 B.C.-A.D. 24) advocated Lao Zi's "non-action thought", which gained ground in the middle Warring States Period. To solidify the centralized feudal autocracy, however, Emperor Wu Di adopted Dong Zhongshu's proposal for honouring the doctrine of Confucius. How to preach and interpret Confucian texts constituted an important aspect of the academic thinking of the Western and Eastern Han periods. That is what historians mean when they use the term "Confucianism of the two Hans". The Confucian School with Dong Zhongshu as its representative was a new-type Confucianism derived from original Confucianism. With the pre-Qin Confucian thinking as the principal content, he absorbed the thoughts of the Legalist School and the Yin-Yang Theory of the Five Elements (wood, fire, earth, metal and water) to establish a theistic thinking characterized by the "correlation between Heaven and man". Sima Qian, a contemporary of Dong who was critical of Dong's metaphysical thinking, believed in a progressive, developmental concept of history. His voluminous work *Shi Ji* (*Records of a Historian*) records more than 3,000 years of Chinese history from the legendary Emperor Huang Di to Emperor Wu Di of the Han Dynasty. It covers, comprehensively, all aspects of Chinese social life and its developments. It has exerted a far-reaching influence on Chinese culture in general and Chinese historiography in particular. Wang Chong of the Eastern Han (A.D. 25-220) in his book *Lun Heng* (*Discourses Weighed in the Balance*) made a penetrating critique of the Confucian thought after Dong Zhongshu, thus elevating materialist thinking of ancient China to a new height.

During the Wei, Jin and the Southern and Northern Dynasties periods (220-581), the Xuan Xue School (Mysterious Learning or Metaphysical School) arose to overshadow Confucianism in popularity. Xuan Xue derived its name from the study of three books: *The Book of Lao Zi, The Book of Zhuang Zi* and *The Book of Changes,* jointly known as the Three Xuans. Actually, followers of the Xuan Xue School used the Taoist thinking to interpret the Confucian classic *The Book of Changes.* They advocated the Taoist philosophy of non-action while upholding Confucian values. It is a sophisticated idealism. Later, Xuan Xue gradually merged with Buddhism that came to China towards the end of the Eastern Han. Fan Zhen launched a resolute fight against the religious superstition of Buddhism. His essay *Shen Mie Lun* (*On the Extinction of the Soul*) was a landmark in the development of classical Chinese materialism. During this period, literary and artistic creation flourished; in keeping with this development appeared literary summaries. Liu Xie, in his book *Wen Xin Diao Long,* (*Carving a Dragon at the Core of Literature*), criticized the trend of formalism that was then in vogue. He summed up the experience in literary creation, and the book was the first systematic work on literary theory in Chinese history.

During the Sui, Tang and the Five Dynasties periods (581-960), Buddhism had its golden age in China, thanks to the support given by the ruling class. With Buddhist ascendancy appeared numerous Buddhist sects preaching different doctrines. Academic thought of this period was closely related to Buddhism, in one way or another. Among those who upheld Confucian orthodoxy and opposed Buddhism was Han Yu, while Liu Zongyuan, Liu Yuxi and others held a materialist atheist view. Historiography also registered a marked progress in this period. Liu Zhiji's *Shi Tong* (*A Chronology*) was China's first progressive and highly critical work on historiography.

In the Northern Song period (960-1126), Wang Anshi's New Learning School advocated the conduct of reforms in order to lessen the acute social contradictions of the time. But his New Learning was vehemently opposed by the "Old Party" among the ruling class. Li Xue, the School of Principles or Neo-Confucianism, came into being as the opposite to Wang Anshi's New Learning. It viewed *li* or "heavenly principles" as the fundamental issue underlining both society and the universe. It was a Confucian philosophy blended with Buddhist and Taoist thoughts. Li Xue prevailed in the Song, Yuan and Ming dynasties (960-1644). It then split into many factions. Among leaders of the idealist faction were Cheng Hao, Cheng Yi, Zhu Xi, Lu Jiuyuan and Wang Yangming, while the materialist faction was represented by Zhang Zai, Chen Liang and Ye Shi. This period also saw masterpieces of history written in a new style, such as Sima Guang's *Zi Zhi Tong Jian* (*Mirror of History*) which covers a period of more than 1,300 years from the Warring States to the Five Dynasties. Although the work lacks the viewpoint of historical development, it was written in a compact, well-organized style rich with material, and it has exercised much influence on the development of Chinese historiography.

Towards the end of the feudal society in China, a large number of learned, progressive thinkers emerged, such as Huang Zongxi, Gu Yanwu and Wang Fuzhi who, under the impact of the peasant wars late in the Ming Dynasty and the overthrow of the Ming Dynasty by the Qing, freed their thinking, in some respects, from the fetters of conventional thought and imbued themselves with new ideas. The feudal rulers of the Qing Dynasty (1644-1911) exercised a rigid thought control and resorted to all measures to turn intellectuals' attention away from social issues. This gave rise to the so-called Han Xue, or Han Learning, that centred on the investigation and annotation of ancient Chinese classics. But people like Dai Zhen and Zhang Xuecheng of the mid-Qing period, following the fine tradition of the late-Ming thinkers, made positive contributions to the academic thinking of the time.

A review of the 2,000-year history of Chinese academic thinking sheds light on two salient features — the tradition of rich materialist and dialectical thinking and the tradition of emphasis on reality and on solution of actual social issues. A dual tradition of this kind has a far-reaching effect on the development of Chinese culture. Particularly, philosophy, historiography and literary criticism attained a high degree of growth in ancient China. They had their own salient features and styles, shining brilliantly in the history of the world and making a valuable contribution to the cultural development of mankind.

However, Chinese academic thinking, which had a brilliant past, began to lag behind during the past few centuries when modern scientific thinking emerged in Europe. The Opium War of 1840 broke open the door of the isolationist, stagnant feudal empire of China. In the face of aggression by foreign capitalist forces, China degenerated from an old feudal society into a semi-feudal, semi-colonial society — which marked the beginning of China's modern history.

How to free China from poverty and backwardness, humiliation and oppression, became an important issue to the Chinese people and intellectuals under the new historical circumstances. Even before the Opium War, some far-sighted intellectuals, who had broken away from the ranks of feudal scholars, sensed the change of times. They desired to free themselves from the fetters of conventional academic thoughts, learn to know the world and carry out reforms. Their representatives were Lin Zexu, Gong Zizhen and Wei Yuan. After China's defeat in the Opium War, Chinese peasants rose in an earth-shaking revolutionary struggle under the banner of the Taiping Heavenly Kingdom. Unfortunately, the struggle was ruthlessly put down by the joint forces of feudal rulers and foreign aggressors. After the defeat, some Chinese intellectuals proposed to emulate the West not only in science but also certain socio-political measures. Learning about Western science and promoting reforms soon became a new trend of thought and a new movement. The trend gave rise to the 1898 movement for constitutional reform and modernization. But that movement, too, ended in failure. In the first decade of this century, the bourgeois revolutionary movement arose and culminated in the 1911 Revolution. That revolution succeeded in ending the political form of monarchal dictatorship, but it did not put an end to the imperialist-feudalist rule. It was in this setting that modern Chinese social sciences appeared. Among the modern Chinese social scientists, there were reformists like Kang Youwei, Liang Qichao, Tan Sitong and Yan Fu. Yan Fu, in particular, exerted an important influence on Chinese social sciences by introducing Western bourgeois theories into China. Dr. Sun Yat-sen, the great forerunner of the bourgeois democratic revolution, formulated the Three People's Principles (Nationalism, Democracy and People's Livelihood) which served as the ideological basis for the revolu-

tionary democracy of his time. Despite their defects, the
Three People's Principles were a most positive accomplishment during the period of the old democratic revolution of China. Zhang Binglin in his early years played
an important role in spreading revolutionary thinking;
after the 1911 Revolution, however, his revolutionary
zeal abated and in his late years he back-slid into the
small world of Confucian classics. Wang Guowei was
politically conservative, but he did some significant work
by applying the methods of Western social sciences to the
study of history. Following the May 4 Movement of
1919, a new world outlook and new methodology came to
China. This new world outlook and methodology guided
by Marxism brought about a change to the academic
world of China. Due to the weakness of the Chinese
bourgeoisie, their social sciences as a whole did not make
much progress; nor did they establish a complete, independent system, with the exception, perhaps, of the
study of historiography and linguistics. Research institutions of modern social sciences appeared even later. For
research in social sciences, the Central Academy of Research which the Kuomintang established in 1928 had
only an institute of history and languages and an institute
of sociology. The Beiping Academy of Research established in 1929 had only an institute of history for the
study of social sciences. There were only a few research
fellows whose limited output had little effect on society.

A New Look of Social Sciences China's social
sciences with Marxism as their guide came into being to
meet the needs of the communist movement during the
May 4 Movement. Li Dazhao, Lu Xun, Guo Moruo and
a large number of other Marxist social scientists made
important contributions to starting and developing this

field of study. Since its very inception, this new field of study has been closely integrated with the Chinese people's liberation movement under the leadership of the Chinese Communist Party. It played a positive role in all the stages of struggle of the new democratic revolution. In the long and arduous revolutionary struggle, Mao Zedong, the great leader of the Chinese people's revolution, painstakingly studied Marxism and the actual problems of the Chinese revolution and ultimately accomplished the most important task of the country's social sciences — creatively applying Marxist theories to the solution of a series of problems confronting the Chinese revolution. He synthetized the wisdom of the whole Party that manifested during the revolutionary struggle and established what has been known as Mao Zedong Thought. This is a great achievement of Chinese social sciences. It is under the guidance of Mao Zedong Thought that the Chinese revolution triumphed.

After the founding of the People's Republic of China, social sciences in China were given an attention unknown in the past. Academic organizations mushroomed, such as the Chinese Society of History, the Chinese Society of Economics and the Chinese Society of Philosophy. In 1950, three research institutes were established under the Chinese Academy of Sciences, in the fields of archaeology, linguistics and modern history. By 1953, more research institutes of social sciences had been established, such as philosophy, economics, literature, history and languages of minority nationalities (the last-mentioned was later merged with the Institute of Nationalities) and an information research office (later changed into the Institute of Information). Between 1958 and 1964, six more research institutes were set up, covering law, nationalities, world economy, world history, world religions and world literature. Under the Department of Philosophy and Social Sciences of the Chinese Academy of Sciences, there were altogether 14 research institutes of social sciences which published a dozen academic journals such as *New Construction*, *Economic Research*, *Law Research*, *Research on Nationalities*, *Philosophical Studies*, *Historical Research*, *Bulletin of Archaeology*, *Archaeology*, *Literary Review*, and *Chinese Language*. Provinces, municipalities and autonomous regions set up 37 research organs of social sciences of their own. Many universities and colleges also established their own social science research organizations. More than 20 academic journals of social sciences, either specialized or comprehensive, were published locally. From then on, researchers of social sciences formed a corps of their own.

The development of social sciences in the People's Republic of China reflected the country's need in the socialist construction and in its international exchanges. For example, the victory of the revolution fired the people's enthusiasm in the study of Marxism, history of Chinese revolution and history of China. It also enhanced the study of Marxist theory and world as well as Chinese history. The progress of the country's economic construction made the study of economic questions ever more pressing. In the course of proceeding with the country's capital construction, a large number of relics and ancient sites were unearthed; archaeological research and the work of museums benefited greatly as a result. The progress of work in the autonomous regions and areas where people of minority nationalities live in compact communities helped advance the study of the societies of minority nationalities, including their history and languages. To disseminate education and culture, Chinese linguists did much work in reforming the Chinese written language, standardizing the Han spoken language and studying the dialects of various localities. In the decade following the founding of the People's Republic, Chinese social scientists translated and published a large amount of works by Marx, Engels, Lenin, Stalin and Mao Zedong. Basing on their intensive studies, they wrote a great deal to popularize Marxist teachings. They also edited a number of ancient Chinese books of great value. They gathered information on many special subjects and introduced to China a number of important works on Western social sciences. They wrote more than 100 textbooks for college students and published many valuable papers on subjects of literature, history, philosophy and economics. Many social scientists went deep in society to study sociological phenomena and dialects. Particularly worth mentioning is the investigation of society, history and languages of minority nationalities, an investigation that was conducted on such a scale and yielded such a wealth of information that hardly a parallel could be found anywhere in the rest of the world. In short, despite the fact that Chinese research in social sciences was still in its infant stage at the time, it achieved a breadth and depth unheard of before Liberation.

There were of course "Left" errors and shortcomings in the study of social sciences before the "cultural revolution". Owing to those errors and shortcomings, the country's social sciences did not make the kind of progress as had been expected. And things got even worse in the ten years between 1966 and 1976 when the Lin Biao and Jiang Qing counter-revolutionary cliques unbridledly advocated modern fetishism which smothered

scientific thinking. They ruthlessly persecuted scientific
workers and dissolved scientific research institutions,
causing scientific research to stagnate and bringing social
sciences to the brink of extinction. It was only after the
ₛshing of the Jiang Qing clique in October 1976 did the
country's social sciences get out of danger.

Beginning of a New Phase of Development   A new
political situation, one of stability and unity, appeared in
China after 1977. Like the national economy, the country's work in social sciences also recovered and grew. The
Third Plenary Session of the 11th Central Committee of
the Chinese Communist Party, held in December 1978,
decided to focus the whole country's work on socialist
modernization. The study of social sciences, too, entered
upon a new phase of development.

In 1977, the Department of Philosophy and Social
Sciences of the Chinese Academy of Sciences was upgrad-
ed and became the Chinese Academy of Social Sciences.
The Fifth National People's Congress held in the following year appointed Hu Qiaomu as President of the CASS.
Since the inception of the CASS in 1977, a number of
new research institutes have been established. Affiliated
with the CASS today are 32 institutes, a research office
and a publishing house. By 1980, the number of academic
journals published by the various research institutes
under the CASS had already reached 49. The research
institutes of social sciences under the jurisdiction of provinces, municipalities and autonomous regions have all
reopened and resumed growth. There are now 134
research institutes of social sciences under the jurisdiction of 28 provinces, municipalities and autonomous
regions. Among them Shanghai, Tianjin, Inner Mongolia,
Liaoning, Jilin, Heilongjiang, Jiangsu, Shandong, Shaan-

xi, Gansu, Qinghai, Ningxia, Xinjiang, Henan, Hubei, Hunan, Guangdong, Guangxi, Sichuan, Guizhou and Yunnan have established their own academies of social sciences. Preparations for establishing an academy of social sciences are under way in the Tibet Autonomous Region. Across the country, 73 research institutes and 212 research offices of social sciences are in operation among institutions of higher learning. Research organs like the Institute of Education administered by various professional departments have been restored and strengthened. Academic journals of social sciences, including special issues on social sciences published by colleges now number more than 100 (excluding those published by the Chinese Academy of Social Sciences). Compared with the situation before the "cultural revolution", the number of social scientists in China has greatly increased.

In the 1978-79 period, meetings were held by the various social science departments to map out medium- and long-range plans for scientific research which laid a good foundation for research on social sciences. Those meetings have proved to be very useful.

In recent years, more than 160 academic societies of various branches of social sciences have been founded. They cover not only major branches of the social sciences, but also minor ones, including very new and unfamiliar subjects. Aside from national societies, there are also regional and local ones. All this indicates how rapidly Chinese social sciences have developed, in both breadth and depth.

Academic discussion on social sciences has also been very active in recent years. Every year, more than 100 academic symposia are held. The debate on the criteria for truth which began among the academics in 1978 soon spread to the whole nation, involving hundreds of millions of people. This debate has proved to be lively education in Marxism, playing a positive role in smashing the mind fetters imposed by the Lin Biao and Jiang Qing cliques and in restoring the Communist Party's fine tradition of seeking truth from facts. The discussion among Chinese economists and economic workers on economic laws under socialism, the readjustment and restructuring of the economy has strengthened the ties between the economists and the economic workers and promoted the cause of both theory and practice. Discussion on major theoretical issues conducted by Chinese literary critics and historians has emancipated people's minds and helped the work of research. The symposium in commemoration of the 100th anniversary of the birth of Lu Xun and the symposium in commemoration of the 70th anniversary of the 1911 Revolution held in 1981 have generated extensive repercussions both at home and abroad.

A new perspective has also been opened up for international academic exchanges. In the years between 1978 and 1981, 245 groups of Chinese social scientists totalling 737 people visited foreign countries, and China received 355 groups of 1,282 foreign scientists. The number of foreign scientists visiting China or Chinese scientists visiting foreign countries keep on growing with each passing year.

All this shows that a fundamental change has taken place in the country's social science research since 1976. This change has brought about a new situation which is described as follows:

First, it must be pointed out that having been freed from the mental fetters imposed by the Lin Biao and Jiang Qing cliques, Chinese social scientists have succeeded in eliminating fetishism. Guided by Marxism-Leninism and Mao Zedong Thought, they have moved valiantly forward in integrating the universal truth of Marxism with the new revolutionary practice of modernizing China. More and more social scientists have come to realize that although Marxism is the guiding thought for all our work, it, being a science itself, is also an object of scientific research, subject to tests by practice. It will develop and grow constantly while it is put into practice. This understanding has enlivened tremendously the theoretical study of Marxism, opened up new realms for study, posed new points of views and enhanced the research of various branches of social sciences.

Second, Chinese social scientists, in line with the principle of integrating theory with practice, are actively studying major problems in actual life, from a theoretical as well as practical point of view. This feature is even more pronounced in economic studies. The large-scale economic development has provided a rich source for economic research and a large number of economists are taking part in the investigations organized by governmental departments and enterprises. The investigations have provided a theoretical basis for the current economic readjustment and restructuring and helped produce more rational programmes. Jurists have taken an active part in building a legal system. Scientists engaged in the study of nationalities, sociology and youth behaviour are conducting investigation in real life. Such a close link between Chinese social scientists and socialist construction has no parallel since Liberation.

Third, the policy of "letting a hundred flowers blossom and a hundred schools of thought contend" has been faithfully carried out. It is very inspiring from the point of view of Chinese social scientists. The bitter lesson of the ten disastrous years of the "cultural revolution" has taught them its importance. Now it has been incorporated as part of the Constitution. The numerous academic journals and symposia have provided a broad stage for its implementation. Scientists are encouraged to air their views freely. They are free to hold debates, free to raise questions, free to criticize or refute criticism. They are encouraged to blaze new trails on the basis of solid research and investigation. The new policy has become a powerful impetus to the development of social sciences in China.

Fourth, another new situation that has emerged is the active, positive attitude towards the study of things outside China. Chinese social scientists are not only diligently learning the advanced experience of the social scientists of other socialist countries, but also studying the achievements of those in Western countries and absorbing what is of value to them. The daily growth of China's relations with foreign countries has created not only unprecedentedly good conditions for Chinese scientists engaged in the study of international issues but also important and pressing tasks for them. There has been a constant increase of systematic information, as well as academic papers on foreign situation. This is indeed an exciting phenomenon.

Finally, while the study of reality is being reinforced the study of basic theories and fundamental issues is strengthened. There is an increase of not only scientific personnel but also science library facilities. Research

progress has been made in philosophy, economics, literature, history, archaeology, linguistics, aesthetics, ethics,
logic, natural dialectics, nationalities, religions, etc.
Researches into many subjects vital to the development
of social sciences are being pursued, and heartening progress has been made.

Despite the outstanding achievements, there are defects and problems in social science research. Chinese
social scientists still have to continue their efforts in
emancipating minds, eliminating fetishism and deepening
the studies of Marxist theories, as well as those of practical investigations. The achievements of recent years
should be considered merely a prelude to a greater cause,
inspiring Chinese social scientists to advance valiantly to
win new victories in the eighties.

2. MAIN ACHIEVEMENTS IN VARIOUS SOCIAL SCIENCES DURING THE LAST THIRTY YEARS

Philosophy Since the establishment of the People's
Republic, research workers in philosophy have consistently integrated theory with practice and succeeded well
in studying and probing major philosophical issues that
arose during the course of the revolution and the construction.

In the early post-Liberation days, Chinese philosophers studied and propagated extensively Marxist ideology. They also emphasized the study of the history of
social development and thus established the fundamental
viewpoint of historical materialism. In 1950 and 1952,
Mao Zedong's two brilliant philosophical works, *On
Practice* and *On Contradiction*, were published. Li Da's

An Explanation of "On Practice" and *An Explanation of "On Contradiction"* came about as a result of studying Mao Zedong's philosophical thinking in that period. Ai Siqi's *Historical Materialism: History of Social Development,* Hu Sheng's *How to Think Effectively,* Yang Xianzhen's *Take Materialism and Dialectics Seriously,* Feng Ding's *Plain Truth* and Yu Guangyuan's *Study Marxist Philosophy* were some of the works that played an important role in the dissemination of Marxist philosophy.

Along with the unfolding of socialist construction and socialist transformation, Chinese philosophical workers began to use Marxist philosophy to study and probe laws that governed the development of socialist society. The areas of study included the relations between productive forces and relations of production, between the economic base and the superstructure, and the nature of class contradiction between the proletariat and the bourgeoisie in China. Discussions on those questions, though unable to reach a consensus, did arouse wide attention among Chinese philosophers and helped in making such studies more profound.

Because of the erroneous attitude taken by N. S. Khrushchov towards Stalin, the Chinese Communist Party, in 1956, published *On the Historical Experience of the Dictatorship of the Proletariat* and *More on the Historical Experience of the Dictatorship of the Proletariat,* both of which made a profound historical materialist analysis of the causes of Stalin's errors and the question of personality cult. The two documents had a far-reaching influence both at home and abroad.

In 1957 Mao Zedong published *On the Correct Handling of Contradictions Among the People.* Mao in this work persistently maintained that the law of unity of the opposites is equally applicable in a socialist society. He pointed out that the basic contradictions in a socialist society remain to be those between the productive forces and the relations of production and between the economic base and the superstructure. He also advanced the viewpoint that in a socialist society there are contradictions among the people as well as contradictions between the enemy and ourselves. There are different ways to tackle the two different kinds of contradictions, he continued. His thought on contradictions was an important contribution to the theory of historical materialism.

In this work Mao Zedong also discussed the law governing the development of truth. He pointed out: "What is correct invariably develops in the course of struggle with what is wrong. The true, the good and the beautiful always exist by contrast with the false, the evil and the ugly, and grow in struggle with them. As soon as something erroneous is rejected and a particular truth accepted by mankind, new truths begin to struggle with new errors. Such struggles will never end." (*Selected Works of Mao Zedong,* Foreign Languages Press, Beijing, 1977, Vol. V, p. 409) This line of thinking constitutes a contribution to the Marxist theory of cognition. In accordance with the law governing the development of truth, he proposed the idea of "Let a hundred flowers blossom and a hundred schools of thought contend" for the purpose of making literature and art thrive and science advance. This idea of his has proved to be very helpful in making the country's socialist culture prosper.

In the period between 1958 and the early sixties, Chinese philosophical workers studied and discussed questions involving objective laws and subjective motivation based on the experience and lessons gained in the course of socialist construction. The view that objective laws may be eliminated was extensively criticized. Emphasizing the objectivity of laws, some people maintained that "only when recognition and action are in conformity with objective laws can they be regarded as subjective motivation". Many people disagreed to this view. They pointed out that subjective motivation plays its role to the fullest only on the basis of recognizing and correctly applying objective laws, but a correct understanding becomes perfect step by step and develops only through practice. The discussion helped propagate dialectical materialism, as it refuted the erroneous viewpoints of idealism and metaphysics.

In the history of the development of materialist dialectics, people differed on the question of universality in the identity of certain aspects of a contradiction. Lenin raised the theme that "development is the struggle of opposites" and "development is the unity of opposites". Stalin emphasized *struggle* to the neglect of *unity*, and this emphasis led him to a series of errors in theory and practice. Mao Zedong carried on and further propounded Lenin's dialectical thinking. *A Concise Dictionary of Philosophy*, published in the Soviet Union, criticized the views in Mao Zedong's *On Contradiction* by saying that there cannot be identity between war and peace, the bourgeoisie and the proletariat, and life and death, on the ground that identity cannot be found in two directly opposite situations. The Soviet point of view was discussed and criticized by interested Chinese during the middle fifties and early sixties. While some were of the opinion that thinking and being are not unity of opposites, others maintained that the identity of opposites in a contradiction is a universal law, that all opposite aspects, including thinking and being, erroneous thinking and being, war and peace, the bourgeoisie and the proletariat, and life and death, under certain conditions, are opposite to each other and mutually interdependent. One can be transformed itself into its opposite, and identity exists in all contradictions. A discussion of this kind upheld dialectics, while refuting the erroneous trend of metaphysics.

Early in the sixties, Chinese philosophers further discussed the relationship between the struggle and identity of opposites in a contradiction and the force of motivation in the development of matters. That discussion enhanced the study of materialist dialectics.

Philosophical works published in this period included *Dialectical Materialism and Historical Materialism* by Ai Siqi, *The Marxist Theory of Cognition Is the Theory of Practice* by Wang Ruoshui, and *Historical Dialectics* by Wu Jiang.

During the "cultural revolution" between 1966 and 1976, the counter-revolutionary cliques headed by Lin Biao and Jiang Qing recklessly distorted and tampered with Marxist philosophy. They advocated voluntarism, alleging that in a socialist society, between social consciousness and social being, the superstructure and the economic base, and the relations of production and productive forces, the former always plays a decisive role. They attacked the view that objectivity is primary and subjectivity is secondary; being is primary and thinking is secondary. Criticizing it as "a reactionary metaphysical view", they pit themselves against the very foundation of materialism. They denied the decisive role which the practice of production played in the development of human history and unjustifiably criticized the

theory of productive forces.   They, in fact, denied the
basic viewpoint of historical materialism.

Not until 1976 and years afterwards did Chinese philosophers free themselves from the fetters.   Once again
philosophical studies in the country began to thrive.

In 1978, a nationwide discussion on criteria for truth
was unfolded in the country.   As a result, modern fetishism as advocated by the Lin Biao and Jiang Qing cliques
was repudiated, the mental fetters they imposed on the
whole Party and people were smashed, the authority of
practice over theory was restored, and the materialist line
of cognition — seeking truth in facts — was reaffirmed.
This was indeed a great emancipation of the minds for
the Chinese people, particularly the Chinese philosophical
workers.

Along with an in-depth discussion on the criteria for
truth, Chinese philosophers in recent years have analysed
the general structure of practice from the viewpoint of
its structural elements and their mutual relations.   Such
an analysis has a theoretical significance of great importance to the understanding of practice and its relations with cognition as well as the way to determine
truth with practice as the criterion.   Some philosophical
workers have studied and discussed the relationship between the relativity and absoluteness of truth itself and
the relationship between truth and falsehood.   They
analysed the objective base for the dual nature of truth
and its epistemological causes.   They studied more
thoroughly the Marxist theory of knowledge.

In 1979, after the People's Republic had entered its
30th year, Chinese philosophical workers pondered on
and summarized, from the philosophical point of view,
the experience, both positive and negative, in the coun-

try's political, economic and cultural activities in the past three decades. Meanwhile, they looked into the new situation and new problems that had come about during the modernization drive.

In June 1981, the Sixth Plenary Session of the 11th Central Committee of the Chinese Communist Party adopted the Resolution on Certain Questions in the History of Our Party Since the Founding of the People's Republic of China. The document contains an overall appraisal of Comrade Mao Zedong's historical role and of Mao Zedong Thought as well. It points out how successfully Comrade Mao Zedong applied dialectical materialism and historical materialism to the work of China's proletarian political party, and this success, in turn, gave birth to the stand, viewpoint and method characteristic of the Chinese Communists, all of which are embodied in three basic tenets, namely, seeking truth in facts, mass line and self-reliance and independence. It analyses Mao Zedong Thought and points out the areas where it is a development of Marxist philosophy. It states that Mao Zedong Thought, as a valuable ideological asset, will continue to guide the Party's action for a long time to come. It stresses that Mao Zedong Thought, as a scientific theory formed and tested over a long period of time, should be made distinct from the mistakes Comrade Mao Zedong made in his later years. It is wrong to deny the scientific value of Mao Zedong Thought or to deny its guiding role in our revolution and construction simply because Comrade Mao Zedong himself had made mistakes in his later years. It is wrong, too, to adopt a dogmatic attitude towards his sayings or to regard whatever he said as the unalterable truth which must be mechanically applied. We must apply and develop Mao Zedong

Thought in our new practice so as to ensure the continued progress of our cause along the scientific course of Marxism-Leninism and Mao Zedong Thought. Those views contained in the Resolution summarize the latest results of the studies of Mao Zedong Thought as they have been carried out by the whole Party and Chinese theoreticians in recent years. They serve as a guide to the Chinese philosophers and other interested people to continue to advance; they serve as the correct orientation and scientific method for workers in Chinese philosophy to deepen further their understanding of Marxism-Leninism and Mao Zedong Thought.

Major writings on Marxist philosophy published in recent years include *Philosophy and Enlightening* by Xing Bensi, *Some Questions Concerning Materialist Dialectics* by Yang Chao, *An Outline of Materialist Dialectics* by Li Da, *On the Philosophical Front* by Wang Ruoshui, *Some Major Questions Concerning Marxist Philosophy* by Zhao Fengqi and others, *Practice, Knowledge and Truth* by Xia Zhentao and others, and finally *The Principle of Dialectical Materialism* by Xiao Qian and others.

Since the founding of New China, scholars of Chinese philosophical history have made considerable progress in studying ancient Chinese philosophy by using the basic theory and method of Marxist philosophy. Much evidence indicates clearly that the history of Chinese philosophy has been by and large the history of unity and struggle between materialism and idealism. Taking the materialist viewpoint that "social being determines social consciousness", Chinese philosophers in recent years made an analysis of various philosophical thoughts, criticizing the bourgeois idealistic method employed by Hu Shi and Feng Youlan in their study of the history of Chinese philosophy. This criticism marked the end of the old method and the beginning of "a leap forward" in the study of Chinese philosophical history under the new method. The accomplishment made under the new method is much greater than that under the old method. According to statistics, in addition to some 6,000 academic papers, more than 200 books have been published, including those on special subjects, compilations of ancient works and materials and translations of ancient writings. In the field of general histories, there are *A General History of Chinese Thoughts* by Hou Wailu (in six books, five volumes), *A History of Chinese Philosophy* by Ren Jiyu (four volumes), *A New History of Chinese Philosophy* by Feng Youlan (books one and two already in print). In the field of special histories, the published works include *A Short History of Chinese Materialism* by Zhang Dainian and *A History of Chinese Logic* by Wang Dianji. Also published were a number of important works of the thoughts of ancient Chinese philosophers. *A Critical Biography of Ancient Chinese Philosophers* by Xin Guanjie and others and *A Critical Biography of Modern Chinese Philosophers* by Xin Guanjie and others (altogether nine books in seven volumes) have been in the process of being published since 1980. In the compilation of ancient works, the latest publications include Guo Moruo's *A Collated Version of Guan Zi* and Wang Ming's *The Collated Book of Taiping,* both of which have attained a high academic standard. *A Selection of Referential Materials Concerning the History of Chinese Philosophy,* in nine volumes and compiled by the Research Group of the History of Chinese Philosophy under the Institute of Philosophy of the Chinese Academy of Social Sciences, is a collection of representative writings of the 3,000 of Chinese philosophy.

It is a fine collection accompanied by notes or translations
in present-day Chinese. It has been warmly received by
scholars at home and abroad since publication. The com-
pilers are making revisions and enriching it in preparation
for a second edition.

Chinese philosophers and scientists have worked together to analyse the history of science and latest developments in science and technology in order to enrich and
develop Marxist philosophy. Initial achievements in this
respect include *Biology and Philosophy* by Tong Dizhou,
*The Struggle Between Materialism and Idealism in the
Development of Physics* by Liu Wanhe, *Some Questions
Concerning the Laws Governing the Development of Natural Sciences* by Gong Yuzhi, *Scientific and Philosophical
Treatises* by Cha Ruqiang, and *Practice and Science* by
Liu Shuzi and others. And philosophical analysis of the
transverse branches of science, cybernetics, information
theory and system theory has aroused keen interest among
many Chinese philosophers. Many researchers have included such concepts as structure, system, layer and information as part of studying materialism, as philosophical
categories for researches to enrich and develop the
materialist outlook on matter, movement and time and
space. Some researchers are applying modern science
and technology such as neurophysiology, psychology,
cybernetics, information theory and system theory to the
study of cognition, and the study reveals the nature and
mechanism of man's course of cognition even more con-
cretely.

Extensive work has also been carried out in the study
of philosophy, logic, aesthetics and ethics of foreign countries and marked progress has been made in recent years.
Translations of a number of works in these fields have

been published. Among the publications by Chinese writers are *French Materialist Philosophy in the 18th Century* by Ge Li, *A Critique of Critical Philosophy* by Li Zehou, *A Critique of the Hegelian Theory on Category* by Pi Zhi and Ru Xin, *On Hegel's Philosophy* by Zhang Shiying, *Comments on Contemporary Western Philosophers* compiled by Du Renzhi, *A Crique of Machism* by Chen Yuanhui, *A Crique of Pragmatism* by Chen Yuanhui, *On W. Wundt* by Chen Yuanhui, *A Commentary on the Frankfurt School* by Xu Chongwen, *The Philosophy of Kitaro Nishida* by Liu Jichen, and *A Study on the Indian Philosopher Vivekananka* by Huang Xinchuan. Works on logic published in recent years include *Formal Logic* by Jin Yuelin, *An Introduction to Mathematical and Physical Logic* by Mo Shaokuei, *An Initial Study of the Logic of Modern Chinese Language* by Chen Zongming, and *The Logic of Mo Zi* by Shen Youding. Publications on aesthetics include *The History of Western Aesthetics* by Zhu Guangqian, *An Introduction to Aesthetics* by Wang Chaowen, *A Critique of Idealist Aesthetics* by Cai Yi, *Discussions on Aesthetics* by Li Zehou, and *The Course of Beauty* by Li Zehou. Works on ethics include *Discussions on Moral Questions* by Zhou Yuanbing and *A Guide to the Study of Ethics* by Li Qi.

A number of high-quality works on philosophy written in plain language for laymen have been published. They include *Practice Is the Sole Criterion for Testing Truth* compiled by the Institute of Philosophy of the Chinese Academy of Social Sciences, *What Is Philosophy?* by Li Xiulin, *Learn to Look at Things in an All-Round Way* by Huang Haiping and Song Linfei, *Proceeding from Reality* by Xu Shijie, *A Talk on Communist Morals* by Sun Yang, *An Introduction to Communist Ethics* by Liu

Qilin and *A Simple Reader of Formal Logic* compiled by
Jin Yuelin.   All these works have been warmly received.

Economics   An important feature in the Chinese
economic theory in the past 30 years or so has been its
close connection with China's economic construction. The
theory has grown along with the reality of socialist construction and the accumulation of experience, and it has
been serving as a daily guide and promoting agent for
practice.

During the 17 years prior to the "cultural revolution", economic studies, under the guidance of Marxism,
achieved considerable success in gaining the experience
and lessons for socialist transformation and construction
and also in tackling the new problems that came up in
the actual conduct of the country's economic life.   At the
same time, however, certain "Left" viewpoints that deviated from Marxism came cropping up.

Between 1953 and 1956 there was a heated discussion among Chinese economists on the economic law that
governed China's transitional period.   Through this discussion, a concrete, in-depth analysis was made regarding the forms, characteristics and mutual relations of socialist economic laws, capitalist economic laws and economic laws of small commodity production that then functioned in the Chinese economic life.   The discussion led
to the recognition of the objectivity of economic laws and
the leading role played by socialist economic laws during
the transitional period, especially the basic economic law
of socialism.   This was the first fruitful try in applying
Marxist economic theory to the reality of Chinese
economy.

In April 1956, Mao Zedong made his famous report
"On the Ten Major Relationships" in which he summed

up China's experience in socialist construction and put forth the task of seeking an appropriate road for socialist construction in conformity with the actual conditions of the country. He also put forward the new viewpoint that greater attention should be paid to the expansion of agriculture and light industry to stimulate the growth of heavy industry. In 1957, he emphasized that the interests of the state, the collective and the individual must all be taken into consideration during the process of socialist construction. In 1959 he put forth the view that in planning the national economy, agriculture be the foundation, and the relative importance of agriculture, light industry and heavy industry be arranged in that order. An original idea of this kind has indeed enriched and developed Marxism-Leninism.

In September 1956, Chen Yun, in outlining the structure of China's socialist economy, pointed out that in industry and commerce, state and collective enterprises should be the dominant force, only supplemented by a certain number of enterprises run by individuals. Likewise, planned production must dominate industrial and agricultural production, only supplemented by free production in the light of market changes. On the question of market, the state market must be in dominance, though free market is allowed to function within certain limits and under the guidance of the state. Chen's proposition, a real breakthrough of the highly centralized economic management structure practised for long by the Soviet Union and some other countries, provided people with new ideas in probing for a proper mode of socialist economy. Soon after, he proposed that the scale of economic construction be in proportion to the nation's economic strength, that equal attention be given to people's livelihood and the state's economic construction, that in drawing up plans, arrangements be made to ensure the steady supply of materials as well as financial and credit balances, and that in arranging material supply, the people's well-being has priority over production and production over capital construction. These views not only are of practical significance to the economic construction then and now but have also added new ideas to the socialist section of political economy.

Through summing up the rich experience of the country's socialist transformation, Chinese economists studied and expounded the need for relations of production to conform with the productive forces as well as the specific forms in which this law has actually functioned in China. They also elucidated how the Chinese people, acting in accordance with this objective laws, had adopted various forms of transition in winning victory in the socialist transformation. Xue Muqiao's works are representative of such theoretical accomplishments.

Following the completion of the socialist transformation in 1956, certain questions appeared, such as the correct way to look at the functions of commodity production and the law of value in the socialist economy, and the relations between the law of value and socialist economic laws. Then a heated discussion went on among Chinese economists. In 1958, the question of commodity production and the law of value again became the centre of attention among the economists when attempts to eliminate commodity production and exchange of equal values failed bitterly in the nationwide drive for the establishment of rural people's communes. Mao Zedong advanced the thesis that continued efforts to develop commodity production is a question of primary importance in building socialism and that the law of value is a "great school" in building socialism and communism. Sun Yefang put forth the far-sighted idea that economic plans should be based on the law of value. Yu Guangyuan expounded the inevitability of the existence of commodity and money relations within the socialist ownership by the whole people and its historical role. Many economists expounded the necessity of socialist commodity production, as conflicts of economic interests do exist between individual workers and workers as a collective group under socialism. They pointed out the identity, rather than a relation of relativity, that exists between the role of the law of value and the role of the law of planned development of the national economy. In a national symposium on economic theories held in Shanghai in April 1959, the above question (as well as the question of distribution according to the amount of work performed and piece rate wage) was discussed fully.

In the period between 1955 and 1959, Ma Yinchu advanced his "new theory on population" in which he stated that considering the fast growth of population in China, it was necessary to introduce family planning and birth control as a means to speed up industrialization. His viewpoint touched the very essence of the problem.

Late in the fifties and early in the sixties, efforts were made to rectify the "Left" errors in the economic work and overcome the serious difficulties in the national economy during the 1959-61 period. In that period, Mao Zedong pointed out that it would not work to expropriate the peasants of their properties or to bypass the necessary stages of development, that violation of the objective economic laws would backfire, and that it is necessary to do well in all aspects of the national economy. Liu

Shaoqi proposed that many producer goods might enter circulation as commodities and that the socialist society needed two labour systems and two educational systems. Zhou Enlai pointed out that the majority of Chinese intellectuals were already intellectuals of the labouring people and that science and technology played a vital role in the country's modernization programme. Chen Yun stressed that targets in plans should conform to reality, that a genuine overall balance of the economy be based on the short supplies of materials and that it was impermissible to maintain planned shortages on the question of material supply. Deng Xiaoping suggested that efforts should be made to consolidate and strengthen industrial enterprises and that workers' congresses be held to improve and strengthen management. Zhu De stressed the importance of developing the handicraft industry and a diversified agricultural production. Deng Zihui suggested that a production responsibility system be introduced in agriculture. All these ideas were of tremendous practical and theoretical significance.

Meanwhile, Chinese economists were engaged in the discussion on three major theoretical issues in the socialist economy, namely, the socialist business accounting system, economic results and reproduction. The discussion on the first two questions was in fact a continuation or development of the discussion on commodity production and the law of value. The discussion resulted in better business accounting for better economic results and better understanding of the role of profits under socialism and of the relations between labour costs and the effective result of labour. It was considered necessary, under socialism, to handle correctly the relation between simple reproduction and expanded reproduction (first simple reproduction and then expanded reproduction). The production of consumer goods has a restrictive effect on the realization of expanded reproduction, and expanded reproduction demands not only $IV + M > II\ C$, but also $II\ (C + M - \frac{M}{X}) > I\ (V + \frac{M}{X})$. There is a ceiling, as well as a floor, for both accumulation and consumption in socialist economic development.

In this period, a number of economic theoretical writings of high academic value were published and textbooks on economy compiled for governmental departments. Furthermore, a number of books based on a systematic compilation of both historical and current economic data were published.

However, under the influence of the erroneous "Left" thought in economic construction, "Left" viewpoints that deviated from Marxism were found among Chinese economists before the "cultural revolution". Some people refused to recognize the objectivity of economic laws, believing that after the basic completion of the socialist transformation, the principal contradiction in the socialist economy remained to be that between the socialist and the capitalist road. They labelled as "revisionist" the idea that striving for the best economic result be the main line through the socialist section of political economy. They had a one-sided emphasis on the advantage of a large population, condemning the proposal for birth control as Malthusianist. They characterized material incentive, the principle of material benefit and the piece rate wage as capitalist, while downgrading the role of the law of value. They condemned as heresy the suggestion to enlarge the role of profit in the socialist business accounting and to set price on the basis of production price. They believed that the accounting unit of the collective ownership should be as big as possible and that the level of public ownership should be as high as possible. And the sham socialist economic theory promoted by the Lin Biao and Jiang Qing cliques during the "cultural revolution" was in fact a malignant growth — a systematization of the above-mentioned "Left" views.

Chinese economics suffered grave damage during the "cultural revolution". Lin Biao, Jiang Qing, Kang Sheng and their followers peddled a whole set of sham socialist economic theories with ulterior motives. They refused to recognize the objectivity of economic laws, opposed the "theory of productive forces", denied the validity of socialist commodity production and the principle of distribution according to performance — all of which, alleged they, contributed to the emergence of new bourgeois elements or even a bourgeoisie within the Communist Party. They, indeed, opposed the socialist business accounting and management system, wrongly criticizing it as "putting profit in command". In the later stage of the "cultural revolution", Jiang Qing and Zhang Chunqiao ordered their followers in Shanghai to put together a "socialist political economy" in an attempt to make systematic their absurdities which, supposedly, would serve as a theoretical basis in their desire to usurp Party leadership and the state power.

Because Marxist economics was trampled underfoot and voluntarism in actual work was encouraged, the national economy was increasingly out of balance, the economy further worsened and the national economy as a whole suffered untold damages. Consequently, serious setbacks to the science and practice of economy were inevitable.

After the downfall of the Gang of Four in 1976, Chinese economists took an active part in the work of rectification in the ideological, theoretical and economic fields. The discussion on the thesis that practice is the only criterion for truth gave tremendous impetus to the emancipation of the mind. People began to give serious thoughts to the cause that repeatedly led to errors in China's socialist construction. They began to search for the proper economic mode, or correct road, for China's socialist modernization in conformity with the reality. They came to recognize that rural economic policies and backward management of industry and communications were the two weak links in China's socialist economy. With this understanding, they stepped up their studies. In October 1978, Hu Qiaomu published an important article entitled *Act in Accordance with Economic Laws and Speed up the Modernization Drive* which proposed guidelines in restructuring China's economic management.

In April 1979, the Chinese economists held a forum in Wuxi, Jiangsu Province, and discussed the role of the law of value in a socialist economy. The forum made an in-depth study of the objective necessity of commodity production and commodity exchange in a socialist society, reviewed the relations between planning and market, and expounded on the necessity of making extensive use of market mechanism in a planned socialist economy. It affirmed the status of socialist enterprises as relatively independent commodity producers, called for attention to the role played by the law of value and the need to implement the principle of material interests, and demanded that production price be the basis for determining prices. All this provided a theoretical foundation for the restructuring of the management system in China's national economy.

In June 1979, the Financial and Economic Commission under the State Council organized economists and other economic workers to study the economic system and structure, the importing of technology, economic theories and methods. Since then, these people have been active in advancing proposals for the restructuring and readjusting of the economy. Their efforts have not only caused the science of economics to thrive and prosper but also provided guidance and impetus to the socialist modernization drive.

The major questions studied in this period include:

On the question of forms of ownership, it is suggested that the kind of ownership or the relations of production that can best promote the growth of productive forces are the best ones; that at the present stage of economic development in China, multiple levels of the development of productive forces do exist and therefore it is necessary to have multiple levels of ownership to suit the situation; that under the premise that the public ownership of the means of production is in a dominant position, diverse economic sectors and diverse forms of management should be allowed to co-exist; that collective ownership in urban areas is of important significance in the present-stage socialist construction; that the "poor transition" (meaning a transition from a lower to a higher level of public ownership despite the backward stage of productive forces) is a blatant violation of the objective economic laws; that in the light of different conditions, it is necessary to have different reward systems based on output insofar as agriculture is concerned, ease policy restrictions on ownership and appropriately enlarge peasants' private plots, etc.

All these measures have effectively helped readjust the structure of ownership and play an important role in activating the socialist economy, especially the rural economy.

On the question of economic structure and socialist economic mode (actually, this is a question of selecting a specific form of ownership and structure), it is stated that socialist economy, though a planned economy, still maintains commodity and monetary relations and needs to use market mechanism to develop energetically socialist commodity production and exchange. In order to establish an appropriate economic structure to suit the above-mentioned situation, it is considered necessary to change the highly centralized state policy-making system into a policy-making system that integrates the state, the economic units and the individual workers, namely to change a unitary regulation system through planning into an arrangement that allows the market to play its fullest regulating role under the guidance of state planning. In other words, an economic management relying mainly on administrative means by the Party and the government is changed into a management through economic organizations, economic measures and economic regulations. The first step to restructuring is to extend the decision-making power to the enterprises. All this, in various degrees, has helped the work of restructuring the economy.

On the question of economic structure, it is pointed out that economic structure includes not only the various proportions in the national economy but also the various branches of the national economy, the various departments, regions, economic sectors and organizations and the mix of the various aspects of social reproduction, their mutual connections and restrictions. The main question is the proper organization of productive forces which generally determine the form of ownership and the economic structure. Overcoming the imbalance of the economy does not amount to the establishment of a rational economic structure, for a rational economic structure must ensure the fullest play of the country's economic advantages, the attainment of the best economic results and the realization of a most ideal economic cycle for the society. A rational economic structure must also meet the requirement of the general laws that govern the development of productive forces as well as the actual conditions of the country, and there should not be a one-sided emphasis on the development of heavy industry as was the case in the past. In view of China's large population, it is necessary to adopt a more medium-level technology, build more medium-sized and small enterprises and develop more labour-intensive projects and trades. It is necessary to readjust the industrial structure, to control the size of the steel-making and machine-building industries, and to speed up the development of agriculture, textile and other light industries, energy source industry, transport and communications, and building industry as well as cultural, educational, and science and public health activities. All this will have a positive bearing on the current work of readjusting the national economy.

On the question of strategy on economic-social development including the choice of proper orientation and goal for the long-range development of the national economy, it is thought necessary, first of all, to take the purpose of the socialist production as the fundamental point of departure and to arrange the medium- and long-range state plans with the end products in view. We, therefore, must abandon the old road of economic development characterized by a high speed, high accumulation, low efficiency and low level of consumption and find a new road of development which requires relatively low investment and accumulation and brings better returns, ensures proportionate economic development, steady growth, a benign cycle, and greater benefits to the population as a whole. The superiority of the socialist system will be brought into fuller play. Moreover, the strategy on economic-social development must be based on the development of agriculture. These considerations help economic workers think over the economic-social development in the light of long-range goals.

The publication of a number of remarkable theoretical works is indicative of the flourishing of the country's economic studies in recent years. Among those which have exerted broad influence are *China's Socialist Economy* by Xue Muqiao, *Some Theoretical Questions Regarding Socialist Economy* by Sun Yefang, *A Dictionary of Political Economy* compiled by Xu Dixin and others, *Probings into the Socialist Section of Political Economy (I)* by Yu Guangyuan, *Management of Socialist State-owned Industrial Enterprises in China* by Ma Hong and others, and *From Classic Economists to Marx* by Chen Daisun.

Now that the focal point of the nation's work has been shifted to socialist modernization, a great many new problems have arisen in the country's economic life. Study of these new problems calls for the establishment of appropriate subject matters in economics to carry out a higher level of research in agricultural economy, industrial economy, trade economy, the study of finance, banking economy, and the study of business management. New branches of economics have come about, such as comparative economy, production force economy, population study, territorial economy, ecological economy, metering economy, and educational economy.

Meanwhile, academic societies of various branches of economics have been established and a large number of economic journals have come into being. As economic exchanges with foreign countries grow, the study of economic development abroad and foreign economic thinking has made rapid progress. Meanwhile, the study of the history of Chinese economy and Chinese economic thinking has attained a new height.

The study of world economy did not exist as an academic discipline before 1949, although a few Chinese economists did study it. The study began in the fifties. In order to meet the needs for actual work, propaganda and teaching, Chinese economists initiated some study on the economy of foreign countries, subjects that concern the world economy and certain theoretical issues related to the world economy. Some scholars' works in this area were translated into Chinese. However, handicapped by the small number of researchers who were scattered across the country and the less than perfect research structure, the study of world economy in that period was anything but systematic and comprehensive, and the subject matter itself was not properly treated as a special subject of research. Some progress was made in the early sixties before the beginning of the "cultural revolution". For example, Chinese economists collected and compiled a great deal of material related to world economy and studied some major topics on it. Some universities and colleges have prepared teaching outlines on the courses of world economy which they offered. The world economy group of the Beijing Economic Society sponsored a number of lectures and forums on the subject matter. Chinese economists conducted extensive discussions on questions such as: Is world economy an independent subject? What are the objects, tasks and methods of a study of this nature? All this has contributed to the cause of socialist revolution and construction of the country.

The study of world economy has continued to develop since 1976. In 1979, a national conference on planning the study of world economy was held and a Programme (draft) for the Study of World Economy in 1978-85 was adopted. Proceeding from the actual situation and keeping in view China's need for readjusting and restructuring its national economy, Chinese researchers in recent years have made in-depth study in world economy and gained valuable results. They gained much in such areas as the general trend of the world economy in the eighties, the roads and modes of the socialist construction in some countries, the growth of the national economy of the developing countries, the positive and negative experience of foreign countries in their modernization drive, the impoverishment of the proletariat under capitalism, monopolist capitalism by the state and transnational corporations of the post-war time, and economic crisis in the capitalist world. Wuhan University and 11 other universities and colleges jointly wrote and compiled the first volume of a textbook on world economy. Beijing University and the Chinese People's University together compiled *An Introduction to World Economy*. An almanac on world economy was also published.

In order to improve the research in world economy and make it better serve the country's modernization needs, the Institute of World Economy under the Chinese Academy of Social Sciences has been working hard to establish and develop a Marxist world economics. In the past two years, Chinese researchers in this area have made an in-depth study of the objects, tasks, methods and theoretical systems of the Marxist world economics and written a series of papers. The book *An Outline World Economy* compiled jointly by tens of experts and scholars from a dozen organizations is being revised for publication at this moment.

To meet the needs for teaching material on world economy in colleges, the Ministry of Education and the People's Publishing House have agreed to publish a series of textbooks on the subject. In order to disseminate knowledge of world economy, the Chinese Social Science Publishing House has invited specialists to write a series of books entitled Common Knowledge of World Economy.

Literature   Theory of literature and art: Considerable progress has been made in the study of the theoretical aspect of literature and art under the guidance of Marxism and Mao Zedong Thought since the founding of new China. In the fifties and early sixties, there were extensive discussions on the question of realism, the question of combining revolutionary realism and revolutionary romanticism, the question of typicalness, the question of creating new heroic models and the question of thinking in terms of images. The study of the question of realism by Mao Dun, Feng Xuefeng, Shao Quanlin and Cai Yi, He Qifang's discussion on the question of regularity in the history of literature and the necessity of inheriting the literary legacy in a critical way, and Wang Chaowen's elaboration on the laws and special features of artistic creation all have their own characteristics. Writings on literature published in the late fifties and early sixties include *Fundamental Principles on Litera-*

ture compiled by Yi Qun and others and *Outline of Literature* compiled by Cai Yi and others. Both works have been adopted by colleges as teaching materials. Zhu Guangqian's *History of Western Aesthetics* has also drawn wide attention. The study of theory in literature and art has made headway in breaking away from the trammels of dogmatism and vulgar sociology in recent years. Intensive efforts have been made to study the literary and artistic thoughts of the creators of Marxist classics; the theory of realism, truthfulness, typicalness and human nature in literature have also been discussed. Positive results have been obtained in the study of the ideas on literature and art and aesthetics in ancient China. All this has been helpful in raising the level of study in the theory of literature and art in China.

Chinese literature: A large number of articles on the literature of ancient China have been published in the past few years, and the number in 1980 alone was close to 2,000. Many works in book form have also been published. Those of higher quality and greater influence include the three-volume *History of Chinese Literature* edited by the Institute of Literature under the Chinese Academy of Social Sciences, the four-volume *History of Chinese Literature* by You Guo'en, Wang Jisi and Xiao Difei, *History of Development in Chinese Literature* (revised edition) by Liu Dajie, *History of Chinese Literary Criticism* by Liu Dajie, *A Perspective on Literature* by Qian Zhongshu and *On the Creation of "Carving a Dragon at the Core of Literature"* by Wang Yuanhua. A large amount of work has been done in compiling complete and individual collections of ancient literary writings such as *A Collection of Ancient Operas*, *A Collection of Poems in the Jin and Yuan Dynasties*, *A*

Collection of Popular Verses in the Jin and Yuan Dynasties, *Additional Collection of Popular Verses in the Yuan Dynasty* and *A Collection of Articles on Ancient Chinese Operas*. Important changes have been made in the *Complete Poems of the Song Dynasty* based on the original edition, and the *Complete Poems of the Tang Dynasty* has been reprinted. A large number of selections from ancient literary works have been published, including *Selection from the "Book of Songs"* edited by Yu Guanying, *Selection of Tang Poems* (in various editions), *Selection of Tang and Song Poems* (in various editions), *Selection of Song Poems with Explanatory Notes* edited by Qian Zhongshu, *Selection of Du Fu's Poems* (two editions edited separately by Feng Zhi and Xiao Difei) and *Selection of Ancient Essays* edited by Guo Shaoyu. The study of the classical novel *Dream of Red Mansions* has not only made great progress in China, but also become an international learning.

A systematic study of the modern Chinese literature between the Opium War in 1840 and the May 4 Movement in 1919 has begun. The *Draft History of Modern Chinese Literature* compiled by the 1956 class of the Chinese Language Department of Fudan University has been off the press. Complete works and collected essays and poems by a number of influential writers have been compiled and published. Moreover, literary works in large volumes, such as the *Collection of Anti-Aggression Literary Works in Modern China* and *Collected Literary Works in the Late Years of the Qing Dynasty* edited by Ah Ying have come out. Approximately 600 academic papers or more have been published over the past 30 years or more. All this demonstrates the great achievements in the study of modern literature, a subject that came about only after the founding of the People's Republic.

A number of books on the history of literature covering the period between the May 4 Movement and 1949 were published in the fifties. They include *Draft History of New Chinese Literature* by Wang Yao, *Brief History of New Chinese Literature* by Ding Yi and *First Draft of the History of New Chinese Literature* by Liu Shousong. They were followed by *History of Ideological Struggle in Chinese Literature and Art* by the Chinese Language Department of Fudan University. Great progress was made in the study of the influential writers Lu Xun, Guo Moruo, Mao Dun, Ba Jin and Cao Yu and their works, and a large number of essays and monographs were published as a result in the fifties and sixties. A rough count shows that more than 120 treatises have been collected in book forms on the study of Lu Xun. Among them, Feng Xuefeng's "My Memory of Lu Xun" and "The Relationship Between Lu Xun and Russian Literature and the Independent Features of Lu Xun's Creative Work" merit attention. In addition, essays by Tang Tao and Chen Yong were also part of the study. In the middle fifties, the *Complete Works of Lu Xun* with annotations was published for the first time. The "Left" trend that had existed in the study of modern literature for a long time has been criticized in the last few years, and practical analysis of the Left-wing cultural movement in the thirties and of the works by Hu Shi, Zhou Zuoren, Shen Congwen and Yu Dafu has been made on the basis of the historical conditions in that decade. Two versions of the *History of Modern Chinese Literature*, edited by Tang Tao and Lin Zhihao respectively and published in 1980, acquired a new outlook. Great efforts have been made to collect and collate materials. For example, more than a dozen volumes of selected works of modern Chinese literature by the Institute of Literature under the Chinese Academy of Social Sciences have come off the press, and a series of books on the history of modern literary movement in China and works by modern Chinese writers are being compiled.

The study of contemporary literature since the founding of New China is a new branch of learning. Selections of short stories, poems, essays, features, one-act plays and children's books were published every year between 1953 and 1961. Beginning in 1979, the People's Literature Publishing House has put out a multi-volume selection of short stories, poems, essays, features, one-act plays and children's books written during the past 30 years. To summarize the achievements and experience in Chinese literature in the ten years after the founding of New China, the Institute of Literature compiled a *Literature in New China's First Decade* and the Chinese Language Department of the Central China Teachers' College compiled a *Draft History of Contemporary Chinese Literature* in the early sixties. In recent years, a lot of work has been done and a number of essays have been written on the achievements and experience in the development of literature after the founding of New China. In 1980, Beijing University, Fudan University and other institutions of higher learning separately or jointly compiled four works on the history of contemporary literature that systematically summarize the experience and lessons in the literature of New China in the past 30 years. Essays and reviews on the works by contemporary writers have been published every year during the past 30 years, except for the "cultural revolution" of 1966-76. These essays and reviews have dealt with almost all important writers and their works. The study of Zhao Shuli, Liu Qing and Liang Bin and their works, including *Builders of a New Life* and *Keep the Red Flag Flying*, has been very intensive. Mao Dun made an outstanding contribution to the review of contemporary literature and to the discovery and training of young writers.

Extensive efforts have been made to discover, collect, collate and translate the folk literature of various minority nationalities in a systematic way. There has been great progress in the discovery and collation of the *King Gesar* of the Tibetan nationality, the *Jangariad* and *Gada Mirin* of the Mongolian nationality, the *Manass* of the Kirgiz nationality, the *Ashima* of the Yi nationality, and the *Zhaoshutun* of the Dai nationality.

The Literature of China's Minority Nationalities, a general information book published in 1981, was the first book giving a systematic description of the literature of the 55 minority nationalities. The book, in three volumes, has about 1.2 million Chinese characters, and its publication represents a most gratifying achievement in the study of the literature of the minority nationalities in recent years. Moreover, a number of works dealing with the history of literature of individual nationalities were published in different parts of China. They include *A Brief History of Mongolian Literature*, *A History of Tibetan Literature*, *A History of Miao Literature*, *A History of Bai Literature*, *A History of Naxi Literature,* and *A History of Zhuang Literature*. All of them give a systematic account of literary tradition of minority nationalities.

Foreign literature: Foreign literary works were translated and introduced to China in an organized and systematic way in the 17 years prior to the "cultural revolution". In the fifties, more attention was paid to the translation of proletarian revolutionary works and the well-known classics of Russia, Britain, France and Germany. In the sixties, the anti-imperialist and anti-colonial literature of Asian, African and Latin American countries was gradually introduced. At the same time, works on Marxist theories of literature and art and histories of literature in the Soviet Union, Britain and France were published.

China began to publish the "Series of Classical Literary Works in Foreign Countries", "Series of Classical Theories of Literature and Art in Foreign Countries" and "Series of Marxist Theories of Literature and Art" after the late fifties.

During this period, emphasis was laid on two aspects of foreign work translations. One was to translate literary works related to the socialist revolution and construction of this period. The translations of *Mother*, *The Making of a Hero* and *Young Guards* were recommended to the young people in China and helped to inspire and educate them. The other was to study famous classical works, such as the works by Shakespeare, the comedies by Moliere and Diderot's aesthetics. In the early sixties, discussions were held on the novel *Le Rouge et le Noir* (*Scarlet and Black*) by Stendhal and on the relationship between Balzac's world outlook and his method of creativity. There were also a number of articles discussing the right attitude towards Western humanitarianism and the way of critically assimilating foreign literature.

Results came from systematic research conducted during this period. The *History of European Literature* (Vol. I) by Yang Zhouhan, Wu Dayuan and Zhao Luorui, the *Brief History of German Literature* by Feng Zhi and the *History of Sanskrit Literature* by Jin Kemu, were published.

The translation of foreign literature was resumed in 1976 after ten years of interruption and has flourished since then. Publishing houses have begun publishing multi-volume selections of works by famous writers and filling the gaps in the translated foreign literature which until then had only consisted of classics or modern progressive literature. Works by modern writers of different schools are also being published. This has broadened the horizons of Chinese readership and exerted constructive influence on Chinese writers.

There have been good results in the study of foreign literature in the short period of past four or five years. The publication of various kinds on the history of foreign literature shows a more thorough and systematic study. Among the published works are the *History of European Literature* (Vol. II) by Yang Zhouhan, Wu Dayuan and Zhao Luorui, *Brief History of Foreign Literature* (European and American part) by Zhu Weizhi and Zhao Li, *History of Foreign Literature* edited jointly by 24 universities and colleges, *History of European and American Literature* by Shi Pu, the three-volume *History of French Literature* (Vol. I) by Liu Mingjiu, Zheng Kelu and Zhang Yinglun and the *Brief History of American Literature* (Vol. I) by Dong Hengxun, Zhu Hong, Li Wenjun, Shi Xianrong and Zheng Tusheng.

The number of essays and treatises on foreign literature has also increased. They include the "First Discussion on *Ramayana*" by Ji Xianling, "On Romain Rolland" by Luo Dagang, "My Impression on Bertolt Brecht's Dramas" by Bian Zhilin, the reprinted "Critical Biography of Gustave Flaubert" by Li Jianwu, the "Critical Biography of Sandor Petofi" by Xing Wansheng, and collected essays and reviews on foreign literature by Wang Zuoliang, Fan Cunzhong, Yang Jiang, Liu Mingjiu and Cheng Daixi.

Meanwhile, there is a "Series of Research Materials on Foreign Literature" devoted to the compilation of large works. Among those already published are *Collection of Reviews on Shakespeare*, *A Study of Ernest Hemingway*, *Collection of Reviews on William Faulkner*, *Collection of Reviews on Charles Dickens*, *Collection of G. Lukace's Essays on Literature*, *Foreign Theorists and Writers on Thinking in Terms of Images*, *European and American Classic Writers on Realism and Romanticism* and *The Question of Realism in Aesthetics Among Russia's Revolutionary Democrats*.

Popular readings of foreign literature include *Biographies of Famous Foreign Writers* by Zhang Yinglun and others and *Summaries of Foreign Literary Works* by Zheng Kelu and others.

The scope of foreign literature research has been extended to cover such countries as Iceland and Sweden for the first time. Growing interest is being shown in comparative literature. Importance has been attached to the introduction and studying of current status in foreign literature.

Historiography Ancient history: Chinese historians pay great attention to the editing, study, and textual review of ancient historical books. In the past 30 years and more, many rare, extremely valuable ancient books have been printed by the photostat process, and not a few historical documents have been edited and published. To make reading a little easier for both Chinese and foreign historians, a number of scholars have got together and completed, in a dozen years, the process of punctuation and checking of the *Twenty-four Dynastic Histories* which has a total of 3,249 volumes and contains 40 million Chinese characters. All of these volumes have been published with punctuations. *A Draft History of the Qing Dynasty*, covering the last dynasty in China, has also been punctuated and published. Publication of these works benefits greatly the study of Chinese history.

More than 100,000 pieces of tortoise shells and animal bones with oracle inscriptions of the Shang Dynasty have been discovered. After long years of hard work, Chinese historians have completed the compilation of a *Collection of Inscriptions on Oracle Bones and Tortoise Shells* and published the book in one volume after another. In recent years, thousands of bamboo and wood inscriptions of the Han Dynasty and large numbers of ancient books have been unearthed in northwest China and other parts of the country. In both quantity and quality, the discovery had never been equalled. The bamboo inscriptions of the Qin Dynasty found in Yunmeng, Hubei Province, the silk inscriptions found in Mawangdui, Hunan Province and the ancient documents unearthed in Turpan, Xinjiang, have attracted the attention of both Chinese and foreign scholars. They are now being compiled and published.

Tens of thousands of national and local documents of the Ming and Qing dynasties are being catalogued and published, step by step, under a long-term plan. Historians are now compiling, annotating and collating ancient historical documents, and compilations and indices have been published one after another. After Liberation, many scholars took part in the investigation of the social life and historical developments of minority nationalities, and investigation reports and brief histories of these nationalities totalling millions of Chinese characters have been written. The materials collected through investigations provide a wealth of knowledge for the study of the history and languages of ancient China. Specialists are now making an in-depth study of these materials and compiling them for publication.

There have been satisfactory results in the multi-purpose as well as special studies of the ancient history of China. Before the "cultural revolution", in addition to a large number of middle-aged and young historians, veterans like Guo Moruo, Fan Wenlan, Jian Bozan, Lu Zhenyu, Hou Wailu, Deng Tuo, Wu Han, Chen Yinke, Chen Yuan and Gu Jiegang wrote many books, articles and monographs and published them. General histories published in recent years include *Draft of Chinese History* (the 1st, 2nd, 3rd and 4th volumes edited by Guo Moruo), *General History of China* (the first six volumes edited by Fan Wenlan and others), *An Outline of Chinese History* (four volumes edited by Jian Bozan), and various versions of *Ancient History of China* compiled by universities and colleges. As for histories of special subjects, there are ideological, philosophical, and economic histories of China. In addition, *History of Peasant Wars in China* and *History of Historiography* have come off the press. New and important works dealing with specific periods in Chinese history, such as the Spring and Autumn Period, the Warring States Period, the Qin and Han dynasties, the Kingdom of Wei, the Jin Dynasty, the Southern and Northern Dynasties, the dynasties of Sui and Tang, the Five Dynasties and the Song and Ming dynasties, have been published. The study of the Qing Dynasty history has been emphasized in recent years. Published works in this area include *Brief History of the Qing Dynasty*, *A Concise History of the Qing Dynasty* and *Founding of the Qing Dynasty*. As for the Qing archives, there are *Series of Historical Documents of the Qing Dynasty*, *Selections from the Confucian Archives in Qufu* and *Historical Data of the Qing Dynasty*. In historical geography, the compilation and study of ancient atlas have been completed. The 8-volume *Collection of Chinese Historical Atlases* edited by Tan Qixiang has come off the press. *A Collection of Maps in Chinese History* (Vol. I) is also available on the market.

Much attention has been paid to the compilation and publication of works by deceased Chinese historians. Published works include *A Collection of Fan Wenlan's Historical Reviews*, *A Collection of Chen Yuan's Academic Papers* (Vol. I), *A Collection of Chen Yuan's Essays on Historical Sources*, and *Collected Works of Chen Yinke*. Works and posthumous manuscripts by Guo Moruo, Gu Jiegang and Lu Zhenyu are being edited and will be published.

Chinese historians attach great importance to not only archaeological discoveries and compilation of historical data but also the exploration of the theoretical aspects of history. Obviously, without a profound and sound theoretical analysis, it would be impossible to replace the outdated traditional historical viewpoints by the scientific interpretation of Chinese history, much less to eradicate the historical prejudice held by the imperialists. Chinese historians have discussed and studied the following historical questions: origin of man in China, definition of the Stone Age, origin of classes and state in ancient China, history of the Xia Dynasty, basic features of the slave system in ancient China, periodization of the slave society and feudal society (including discussion on the social nature of the Western Zhou Dynasty to the Han Dynasty), origin of the Han nationality, characteristics and historical role of the feudal autocracy in China, formation and characteristics of feudal land ownership system in China, causes of the prolonged continuation of the feudal society in China, nature of the peasant wars in China and their characteristics and historical role, evolution and historical impact of Confucian thought, budding of capitalism and disintegration of the feudal system in China, forces of motivity in historical development, Asiatic mode of production, and relationships among Chinese nationalities. A scientific interpretation of the long history of feudalism in China will help the Chinese people of today to appreciate and carry on their own fine cultural tradition and to eradicate the undesirable, bad influence of backward feudalism, thus enabling them to achieve modernization at an earlier date. Scholars of China's ancient history are working hard to attain this goal.

Modern history: As for the modern history of China, the erroneous view that imperialist aggression had brought China "civilization" and "progress" had considerable influence before Liberation. Since the founding of New China, the historians, in an effort to eradicate the old influence, have applied Marxist viewpoint to the study of China's modern history and singled out the contradiction between the Chinese people and imperialism and feudalism as a key issue in the modern history of China. Many academic papers were written under the guidance of this basic viewpoint.

About a dozen general histories of the modern period have been published in the past 30 years or so. Among the published works is *Modern History of China* (Vol. I, revised edition of 1951) by Fan Wenlan, which has enjoyed a large readership. The fourth volume of *The Draft of Chinese History* by Guo Moruo and the fourth volume of the *Outline of Chinese History* by Jian Bozan, both published in the early sixties, are special volumes covering the history of modern China. The first volume of a three-volume book entitled *Draft of China's Modern History* by Liu Danian has come off the press. Hu Sheng's *From the Opium War to the May 4 Movement*, in two volumes, has also been published. The first volume of the *Draft of Modern Chinese History* by Dai Yi was published in the late fifties, but the other volumes have not yet been in print. The *History of the New Democratic Revolution in China*, written by Li Xin and others, covers the latter part of China's modern history, namely the period between 1919 and 1949.

There are other books dealing with a special aspect or subject of China's modern history. Among books about imperialist aggression against China are *History of U.S. Aggression in China* by Liu Danian, *History of U.S. Aggression in China* (Vols. I and II) by Qing Ruji and the first volume of *History of Imperialist Aggression in China* by Ding Mingnan and others, all published in the fifties. Recent books on the same subject (some republished after revision) include the first two volumes of *History of Tzarist Aggression in China* by Yu Shengwu and others, *Imperialism and Chinese Politics* by Hu Sheng, and *Chinese Politics and the 1911 Revolution* by Li Shu. Books that cover the major events of modern China include *A Draft History of the Taiping Heavenly Kingdom* by Luo Ergang, the three-volume *Draft History of the 1911 Revolution* by Jin Chongji and Hu Shengwu and the three-volume *History of the 1911 Revolution* by

Zhang Kaiyuan and Lin Zhengping. The research on the history of the Republic of China has been developing in recent years, and publications include the *Collection of Historical Materials of the Republic of China* of various editions, the first three volumes of *Biographies in the Republic of China* by Li Xin and Sun Sibai and *The Chronicle of Dr. Sun Yat-sen* by the Institute of Philosophy and Social Sciences in Guangdong Province. The first volume of *The History of the Republic of China* by Li Xin has come off the press.

About 5,000 articles on the various aspects of China's modern history have been published in the Chinese press in the past 30 years. Many of these articles touched upon important questions and generated much discussion. These questions concerned the periodization of modern history, character of the Taiping Revolution and process of Hong Xiuquan's ideological development, character and role of the Westernization Movement, character and role of the Yi He Tuan Movement of 1900, role of the bourgeoisie in the 1911 Revolution, etc. In recent years there has been a debate on the basic "link" in the history of modern China.

Historians have gathered a large quantity of materials on modern China in the past 30 years. During the early years after Liberation, the Chinese Society of Historians began to compile a "Modern Chinese History Materials Series" and later published them one after another. The series covered the Opium War, the Second Opium War, the Taiping Heavenly Kingdom, the Nian Army (Torch Bearers), the Hui Uprising, the Westernization Movement, the Sino-French War, the Sino-Japanese War, the Reform Movement of 1898, the Yi He Tuan Movement and the 1911 Revolution, totalling more than 20 million Chinese characters. A series of reference books on the modern economic history of China compiled by the Institute of Economics under the aegis of the Chinese Academy of Social Sciences include *Selected Statistics on the Modern History of Chinese Economy, Source Materials on the History of China's Modern Industry, Source Materials on the History of Agriculture in Modern China, Source Materials on the History of Handicrafts in Modern China, Source Materials on the History of Modern China's Foreign Debts, Source Materials on the History of Chinese Railways* and *Source Materials on the History of China's Foreign Trade.* They total about six million Chinese characters. The series of 15 books under the title of *Imperialist Powers and China's Customs* are selected from customs archives in old China and *Source Materials on the Reform Movement of 1898* originates in the archives in the Imperial Palace of the Qing Dynasty. The more than 40 issues of the *Source Materials on Modern History* have provided a rich ground for publication.

World history: World history as an independent branch of learning came into existence after the founding of the People's Republic. Before Liberation, China had not a special institution for the study of world history, and only a few scholars taught it in universities and colleges. There was only a limited number of publications on foreign histories. After Liberation, study of world history was emphasized and much work was done in training personnel and collecting materials. A good number of researchers specializing in world history have been brought up.

There have been good results in the study of world history over the past 30 years. For example, a number of books on foreign history have been translated into Chinese, important materials on world history have been compiled, and not a few textbooks on world history have been published. Period histories, histories of different countries or histories dealing with special subjects, plus a four-volume general history of the world, have been published. Published too were many scholarly papers. According to incomplete statistics, about 400 titles on world history have been translated and published, and about 300 original writings have been published in China since Liberation. Chinese scholars have held valuable exploratory discussions on Sino-foreign relations and cultural exchanges and some important questions on the history of foreign countries and presented their own independent views. In August 1980, the Chinese Society of Historians sent a delegation to the 15th International Conference on Historiography in Bucharest where the delegates presented papers and exchanged views with their foreign colleagues.

Today, Chinese scholars in world history are studying more and more subjects in a much wider scope. However, judged by accepted international standard, Chinese expertise in world history is still very weak, the number of historians in this field is still very small, and there are not many truly accomplished. A good number of sub-branches, such as history of historiography, history of the Hittites, history of Byzantine, history of papyrus and history of metrology are virtually blank; there are many difficulties in this respect. Chinese scholars in world history are in the process of overcoming these shortcomings and working hard to make their own contribution to the study of world history.

Archaeology: As an academic discipline, archaeology has done well in the past 30 years or more. Archaeological

investigations and discoveries have been made in all parts
of China, archaeological studies have become more and
more refined, and archaeological achievements have
attracted worldwide attention.

The archaeological discoveries of the Old Stone Age,
apart from the pre-Liberation discoveries of Peking Man
and Hetao Man, also include the discoveries of the even
earlier Lantian Man and Yuanmou Man. Yuanmou Man
lived 1.7 million years ago, and Lantian Man, much
younger, was about 750,000 to 650,000 years old. Old Stone
Age sites, simultaneous with or a little later than the sites
of Peking Man, included Kehe, Dingcun, Shiyu,
Xiaonanhai and Xiachun. Simultaneous with Peking
Man were also the fossils of skull and lower jawbone of
the erect ape-man found in the Longtan Cave, Anhui
Province. Fossils of ape-man, who lived at a later stage
than Peking Man, were also found in such places as Maba,
Changyang, Dali, Liujiang and Ziyang.

About 6,000 or 7,000 sites of the New Stone Age have
been found, and about 100 of them have been excavated.
The large quantity of data thus accumulated have provided a clearer picture of the primitive cultures in different
parts of China and filled many gaps in the geographical
or historical knowledge. The most spectacular of them are
the Cishan Culture and the Peiligang Culture once flourishing in the middle reaches of the Huanghe (Yellow River)
Basin — both of which were earlier than the Yangshao
Culture. The Hemudu Culture along the lower reaches of
the Changjiang (Yangtze River) corresponded to or was
even earlier than the early stage of the Yangshao Culture.
The discoveries of the early stage of New Stone Age cultures are significant and helpful to the study of the origin
and development of New Stone Age cultures in general,

including the origin of agriculture, animal husbandry and ceramics in China. The study of the geographical distribution and types of the Yangshao Culture is progressing in depth. In addition, new cultures of the New Stone Age have been discovered, such as the Jianchuan Culture, the Xiqiaoshan Culture, the Tanshishan Culture, the Daxi Culture, the Qujialing Culture, the Majiabang Culture, the Qingliangang Culture and the Dawenkou Culture. The new discoveries have laid a solid foundation for an in-depth study of primitive cultures in China.

In recent years new clues have been found in the course of studying the Xia Dynasty, the first dynasty in China, but no archaeological evidence has ever been found to indicate its factual existence.

Nevertheless, significant progress has been made in the study of the Yin-Shang cultures. The ruins of an ancient city, dating back to the Yin-Shang Dynasty and earlier than that found in Anyang, was discovered at Erligang near Zhengzhou. The site of a royal palace, built on an earlier foundation of rammed earth (100 × 108 metres), was located at the Erlitou ruins near Yanshi. The Erlitou ruins was believed to be either Xibo, capital of the Shang Dynasty, or ruins of an earlier city flourishing in the later part of the Xia Dynasty. The unearthing of the Yin ruins at Anyang led to many important discoveries. A large tomb was uncovered at the Wuguan Village in the early post-Liberation years. In recent years, more than 4,000 pieces of oracle bones and tortoise shells were discovered, and the tomb of Fu Hao, the spouse of Wu Ding, was also unearthed. The tomb was the first well-preserved tomb of a royal family member of the Shang Dynasty ever unearthed in the past 50 years. Important ruins of the Yin-Shang Dynasty, graves and surviving properties were discovered at Panlongcheng in Huangpi, Hubei Province, Wucheng near Qingjiang in Jiangxi Province, Huangcun in Ningxiang, Hunan Province, Yue'erhe in Funan, Anhui Province, Taixi in Gaocheng, Hebei Province, Qiuwan in Tongshan, Jiangsu Province, Sufutun in Yidu, Shandong Province, Liujiahe in Pinggu, Beijing, and Beidongcun in Kazuo, Liaoning Province. These discoveries have greatly enriched the known content of the Yin-Shang cultures.

The excavation at Fengjing and Gaojing, both capitals of the Western Zhou Dynasty at one time or another, has become a standard, against which the archaeological chronology of the Western Zhou Dynasty is measured. A large number of noblemen's graves, rich in bronze vessels, as well as hoards of bronze, were found in Xian and nearby areas, such as Fengxiang, Qishan and Baoji. Tombs with bronze ware of the Western Zhou Period were also found at Baicaopo in Gansu Province, Beiyao in Loyang, Henan Province, Pingba in Jingshan, Hubei Province, Tunxi in Anhui Province, Yandunshan in Dantu, Jiangsu Province, Huangxian in Shandong Province, Xizhangcun in Yuanshi, Hebei Province, Liulihe in Fangshan and Baifou in Changping, both near Beijing, and Machanggou in Lingyuan, Liaoning Province. These bronzes often have on them important inscriptions, which prove to be highly valuable to the study of the social and historical developments at that time. Remnants of palaces and temples of the early Western Zhou Period have also been found around Fufeng and Zhouyuan in Qishan, both in Shaanxi Province.

The excavation at the Eastern Zhou graveyard on the Zhongzhou Road, Loyang, can likewise serve as a standard, against which the archaeological chronology of the

Eastern Zhou Period in that area can be measured. Many exquisite articles were unearthed from the graves of the Eastern Zhou noblemen, such as the tomb of Crown Prince Guo at Shangcunling, the tomb of Marquis Cai in Shouxian County, the tombs of the Chu noblemen in Jiangling and Xinyang, the tomb of Marquis Zeng in Suixian County and the tomb of Prince Zhongshan in Pingshan. In recent years, the ruins of a copper mine dating back to the Eastern Zhou Period was found at Tonglushan in Hubei Province. Surveying and excavation have been done at the ruins of several capital cities of the Eastern Zhou Period, such as Linzi of the Qi State, Handan of the Zhao State, Xiadu of the Yan State, Jinan of the Chu State, and Houma of Jin and Wei states.

Archaeological discoveries regarding the Qin and Han dynasties have also made headway. The ruins of an imperial palace have been found at Xianyang, capital of the Qin Dynasty. Three pits containing terra-cotta figures have been uncovered at Lintong, Shaanxi, in the general area where the Mausoleum of the First Emperor of the Qin Dynasty is located. The pits contain thousands of terra-cotta soldiers and horses and wooden chariots. Recently, a pit with bronze horses and chariots was discovered. Such discoveries have been unprecedented in China's archaeological history. Moreover, more than 1,000 pieces of bamboo inscriptions were uncovered in the graves of the Qin Dynasty at Shuihudi in Yunmeng County, Hubei Province. The inscriptions narrate mainly the articles of law of that time. The excavation at Changan (known as Xian today) of the Western Han Dynasty and Loyang of the Eastern Han and Wei dynasties has also been very fruitful. Similar excavation of the Han tombs at Mawangdui in Changsha and Lingshan in Man-cheng County produced large numbers of exquisite burial articles. In the Mawangdui tomb was a well-preserved woman corpse. Wood inscriptions recording rites and medication practised during the Han Dynasty were found in a Han tomb in Wuwei; wood inscriptions bearing "Master Sun's Art of War" and "Sun Bin's Art of War" were likewise found at Yinqueshan in Linyi. Along Juyansai, archaeologists also found Han wood inscriptions in the ruins of beacon towers. They are all important discoveries.

Much progress has been made in excavating the ruins of Buddhist structures, ruins of cities and graves of the Wei Kingdom, the Jin Dynasty, the Southern and Northern Dynasties, the Sui Dynasty and the Tang Dynasty. The layouts of the capital of Changan and the temporary capital of Loyang, both of the Sui and Tang dynasties, have been made clear. Important archaeological results were obtained in excavating the tombs of Tang princes and princesses buried with Emperor Taizong, Emperor Gaozong and Empress Wu Zetian of the Tang Dynasty, such as Princes Zhanghuai and Yide and Princess Yongtai, all located in the outskirts of Changan.

The archaeological work on the Song, Yuan and Ming dynasties includes the excavation of Zhongjing of the Liao Dynasty and Dadu of the Yuan Dynasty and the survey of Zhongdu in the early Ming period. Among the important findings are two tombs of the Southern Tang in Nanjing during the Song Dynasty, a tomb of the Song Dynasty at Baisha, Henan Province and a Ming tomb located to the north of Beijing. Excavations at the ancient kilns of Yaozhou in Shaanxi Province, Dehua in Fujian Province and Longquan in Zhejiang Province have all yielded gratifying results.

Important discoveries along the "Silk Road" of the Han and Tang dynasties showing the historical links between China and other countries include silk fabrics, official documents of the Northern Dynasties and the Tang Dynasty, silver coins of the Persian Sassanid regime and gold coins of the East Roman Empire. A shipyard for constructing sea-going vessels and whatever remains of the ships manufactured during that period was also located.

Archaeological accomplishments in the minority nationality regions include important discoveries about the people of Xiongnu and Donghu in north China, the states along the "Silk Road", the Bashu Culture in Sichuan Province and the Dian State Culture in Yunnan Province.

The methods used in natural science have been successfully applied in archaeology discoveries, such as the use of Carbon-14 for dating, thermal release for the division of dynastic periods, chemical and spectrum analysis, and the methods used in physical anthropology and archaeological zoology. During the past 30 years or more, marked progress has also been made in using archaeological data for the study of the history of metallurgy, textile and ceramics.

Linguistics Three major tasks were set at the National Language Reform Conference and at an academic meeting on the standardization of the modern Chinese language, both held in October 1955. The three major tasks were: reform of the Chinese written language, popularization of the standard spoken language based on the Beijing dialect and standardization of the Chinese language. There has also been success in the study of linguistics and philology as it is shown below:

The reform of the Chinese written language: The State Council made public the Scheme for Simplifying Chinese Characters in 1955, the Scheme for the Chinese Phonetic Alphabet in 1956 and the Second Scheme for Simplifying Chinese Characters (draft) in 1978.

Chinese dialects: A general survey of the dialects in nearly 2,000 localities in China was conducted, and many monographs based on the study of local dialects, together with a book entitled *General Description of Dialects*, were published.

Popularization of the standard Chinese spoken language: Three tables of differently enunciated and yet standard pronunciations, together with tables of standard pronunciations for geographical names, were carefully examined and then approved by the Pronunciation Examination Committee for Standard Spoken Chinese. Pamphlets prepared for learning standard spoken Chinese have been prepared for dozens of areas that use different dialects.

Compilation of dictionaries: The following dictionaries have been edited and published: *Modern Chinese Dictionary*, *Pocket Modern Chinese Dictionary*, *Xinhua Dictionary*, *Shorter Xinhua Dictionary*, *Dictionary of Ancient Chinese Words for Everyday Use* and *Dictionary of Chinese Phrases*. The compilation of two large Chinese dictionaries are now in progress, and the *Cihai* (*Dictionary of Words and Phrases*) and *Ciyuan* (*Origins of Words and Phrases*) have both been revised and reprinted. Dictionaries of minority nationality-Han and Han-minority nationality languages as well as foreign-Chinese languages dictionaries have also been published.

Results in the study of languages: Good results have been obtained in the study of linguistic theory, linguistic history, ancient Chinese language, modern Chinese language, history of Chinese language, use of languages, phonetic experiments and modern Chinese language. Published works include *Selected Theses of Luo Changpei on Linguistics, Outlines for Common Phonetics* by Luo Changpei and Wang Jun, *A Guide to Grammar and Rhetorics* by Lu Shuxiang and Zhu Dexi, *800 Modern Chinese Words* by Lu Shuxiang, *Analysis of Chinese Grammar* by Lu Shuxiang, *A Guide to Modern Chinese Grammar* by Ding Shengshu and others, *Handbook on Comparisons Between Ancient and Modern Chinese Characters and Pronunciations* by Ding Shengshu and Li Rong, *A Draft History of the Chinese Language* by Wang Li, *History of Chinese Linguistics* by Wang Li, *Ancient Chinese Language* by Wang Li, *Formation of Chinese Words* by Lu Zhiwei and others, *Brief History of Linguistics* by Cen Qixiang, *On the System of Pronouncing a Chinese Character by Using Two Other Characters* by Li Rong, *Collection of Theses on Chinese Grammar* by Zhu Dexi, *Theory of Grammar* by Gao Mingkai, *Eighteen Lessons on Chinese Grammar* by Li Jinxi, *On the Reform of Chinese Characters* by Zhou Youguang, *Modern Chinese Language* by Hu Yushu, *Essentials of Chinese Dialects* by Yuan Jiahua and *A Revised Edition of "Chinese Pronunciations"* by Zhou Zumo.

Intensified and enlarged efforts in the study of different branches of linguistics: Different aspects of the Chinese languages have been studied, such as the coding of Chinese characters and information processing systems, phonetic experiments, translation by machines, engineering linguistics, mathematics and physics linguistics, physiolinguistics, and psycholinguistics. Study has also been conducted on the relationships between linguistics

and other branches of social sciences, such as socio-
linguistics, children's language, language and thinking,
and language and logic. These studies are still at their infant stage and will be strengthened in the future.

Language teaching: Standardized teaching materials
are provided for courses on linguistics and languages
among the universities, teachers' colleges and institutes
of nationalities. Courses include "Ancient Chinese
Language", "Modern Chinese Language", "Introduction
to Linguistics", "Common Linguistics", "History of the
Chinese Language", "Investigations on Chinese Dialects",
"Theory of Grammar", "Reform of the Written Language",
and "Phonology of the Chinese Language". The book
entitled *Provisional System for Teaching Chinese Grammar in Secondary Schools* has been approved for classroom use.

Moreover, books and reference materials on linguistics published before Liberation have been reprinted;
foreign books on similar subjects have been translated.
During the past 30 years or more, more than 1,000
books and more than 10,000 academic papers on linguistics and philology have been published.

Law Since the founding of the People's Republic
of China, the study of law has made steady progress side
by side with the development of the socialist legal
system.

During the early years after Liberation and during
the period of the first two five-year plans, Chinese jurists
took an active part in the drafting of the Common Programme of the Chinese People's Political Consultative
Conference, Marriage Law of the People's Republic of
China, Land Reform Law of the People's Republic of
China, Trade Union Law of the People's Republic of

China and Regulations of the People's Republic of China
on the Punishment of Counter-Revolutionaries and the
Constitution of the People's Republic of China. At the
same time, research on the science of law began, and
special research departments were established. Universities and colleges in many parts of the country added departments of law or special institutes of political science
and law. Important issues on law were discussed, such
as objects of theoretical study in state and in law, relationship between state and law, nature and role of law
during the socialist period of China, relation between
policy and law, class nature and inheritability of law,
correct distinguishing between and handling of the two
different categories of contradictions, system of people's
congresses, relation between cause and effect in the
criminal law, types and nature of punishment, object in
adjusting the civil law, object and scope in China's history
of legal system and problems of system in international
law. China also published *Teaching Materials on the
Constitution of the People's Republic of China, Teaching
Materials on the Criminal Law of the People's Republic of
China, Teaching Materials on the Civil Law of the People's Republic of China, International Law* by Zhou
Gengsheng, and many other papers.

The establishment of China's legal system entered
into a new phase after 1976. Learning a lesson from past
experience, people began to understand the importance
of strengthening the socialist legal system. In the short
period of a few years since 1976, China has formulated
Organic Law of Local People's Congresses and People's
Governments of the People's Republic of China, Electoral
Law of the National People's Congress and Local People's
Congresses of the People's Republic of China, Organic

Law of People's Courts in the People's Republic of China, Organic Law of People's Procuratorates in the People's Republic of China, Criminal Law of the People's Republic of China, Law of Criminal Procedure of the People's Republic of China, Law of the People's Republic of China on Joint Ventures with Chinese and Foreign Investment, Nationality Law of the People's Republic of China, Marriage Law of the People's Republic of China, Income Tax Law of the People's Republic of China on Joint Ventures with Chinese and Foreign Investment, Personal Income Tax Law of the People's Republic of China, Law of the People's Republic of China on Economic Contracts and Foreign Enterprise Income Tax Law of the People's Republic of China. Chinese jurists took an active part in the drafting of the above-mentioned laws and discussed the important theoretical and practical issues in the establishment of the legal system. The questions cover democracy and legal system, the nature and role of law, equality before the law for all citizens, human rights, the issue of rule by man or by law, judgement of innocence, and the relation between civil laws and economic laws. Among published books on law are *Fundamental Theory of Law* by Chen Shouyi and others, *History of China's Legal System* (Vol. I) by Zhang Jinfan and others, *Fundamentals of International Private Laws*, *Brief History of China's Legal System*, and *Short History of China's Constitutions*. Other published books are: *Selected Materials on Legal Systems During the Democratic Revolutionary Period*, *Selected Economic Laws and Regulations of the People's Republic of China*, *A Collection of Materials on the Criminal Law of the People's Republic of China*, *Collected Materials on the Law of Criminal Procedure of the People's Republic of*

China and *Selected Materials on People's Mediation Work.*
Today there are more research and teaching departments
on law than before the "cultural revolution".

Research on Ethnic Nationalities   The 17 years between 1949 and 1966 marked the first phase of China's
research work on minority nationalities. The major tasks
completed during this period were the differentiation of
various nationalities, the investigations on their social
histories and languages, and the beginning of writing
about their short histories and the short histories of their
languages.

In 1950, the minority nationalities in China were only
recently emancipated from old China's oppression and
prejudice. In order to abolish the oppression practised
by the reactionary government in old China, which ignor-
ed the existence of the minority nationalities, and to adopt
the policy of equality among all nationalities, the Central
People's Government organized interested specialists,
scholars and cadres into groups and sent them to the nationalities' areas to identify them and learn about their life.
In 1956, based on an analysis of societies, histories, languages, and cultures, the total number of historically recognized nationalities, as later announced by the State
Council, was 51. The Pumi nationality in Yunnan and
the Moinba and Lhoba nationalities in Tibet were added
later; so was the nationality of Jino in Yunnan added recently. There are now a total of 55 minority nationalities
in China, or 56 nationalities including the Han.

In 1956, the areas inhabited by minority nationalities
underwent the high tide of democratic reform and socialist
transformation, and all of them experienced profound
changes. Then the urgent task was to preserve the
materials on their original societies, economies, cultures,

habits and customs. The necessary work and investigation began under the leadership of the Nationalities' Committee of the National People's Congress; later, they continued under the leadership of the Nationalities Institute of the Chinese Academy of Sciences established in 1958. Total number of investigators exceeded 1,000 at one time; among them were specialists, teachers and students of universities and colleges, local cadres, and cadres of various nationalities. After the investigation, more than 190 kinds of materials were printed, totalling 14 million Chinese characters; later, about 150 additional kinds of materials were compiled and ready to be printed, totalling 15 million Chinese characters. These materials were later used as primary sources by interested parties to carry out changes and transformation in the areas of minority nationalities. They were also helpful in the study of national characteristics as well as the general law that governed the history of social development of all men.

Films were taken to show the characteristics of a primitive society, slavery and serfdom among some nationalities. They were used to complement and confirm the literary data collected from on-the-spot investigations.

Brief histories of 51 minority nationalities and three series of "General Information on Autonomous Areas" were then written and compiled.

At the same time, the Languages Institute of Minority Nationalities (later merged into the Nationalities Institute), under the aegis of the Chinese Academy of Sciences, was busy making on-the-spot investigations of the languages of the more than 50 minority nationalities. The investigators studied the written languages of some nationalities as they were originally written and learnt about how both their written and spoken languages were used. On the basis of these investigations, they designed either schemes for alphabetical writing or schemes for improving written languages, to be used by some of the minority nationalities.

As the study became more intensive, more questions were also raised and discussed. In the late fifties and early sixties, a number of important symposiums and meetings were held to discuss such questions as formation of nationalities, standard translation of the word "nationality", historical relations among nationalities, and correct appraisal of some historical figures among the minority nationalities. Forums, meetings and special discussions were held to discuss the languages of the minority nationalities.

The second phase in the study of nationalities began in 1977. The task for this period was the continuation of the same study that had been interrupted by the "cultural revolution". The purpose of this study was to lay emphasis on the theoretical and practical problems then encountered concerning the work of nationalities, and the solution of these problems would serve the cause of socialist modernization in areas inhabited by the minority nationalities. At the same time, efforts were made to strengthen the related branch of academic learning and laws of science were strictly observed so as to raise the quality of the research work.

In 1979, entrusted by the Chinese Academy of Social Sciences, the Nationalities Institute called a national meeting to map out a research programme for 1979-85. The research projects in the programme cover the theory and the policy concerning nationalities, the study of nationalities around the world, the history of minority nationalities

in China and the relations among them as recorded in
Chinese history, the differences between various nationalities in China, their forms of social development, cultural
characteristics and process of modernization, and finally,
their languages. More than 500 research items were
listed.

Among the participants in the study are the Nationalities Institute under the Chinese Academy of Social
Sciences, the nationalities research departments in the
provinces and autonomous regions, the Central Institute
of Nationalities, Beijing University and other colleges and
universities, more than 80 units altogether.

Moreover, the State Nationalities' Affairs Commission
has established editorial boards for five series of books
on nationalities. They are the "Minority Nationalities in
China", "General Information on Autonomous Regions
and Areas", "Brief Histories of Minority Nationalities",
"Brief Annals of the Languages of Minority Nationalities"
and "Investigation Materials on the Social Histories of
Minority Nationalities", all of which are major items in
the programme. The Nationalities Institute under the Chinese Academy of Social Sciences, in co-operation with
some other departments, is writing and compiling 21
"Brief Histories of Minority Nationalities", 54 "Brief An-
nals of the Languages of Minority Nationalities" and 26
"Investigation Materials on the Social Histories of
Minority Nationalities". So far, the collaborators have
completed and published the *Short History of the Manchu
Nationality, Short History of the Hui Nationality, Brief
Annals of the Zhuang Language, Brief Annals of
the Bouyei Language, Brief Annals of the Dai
Language, Brief Annals of the Maonan Language,
Brief Annals of the Dong Language, Brief Annals of the*

Mulam Language, *Brief Annals of the Shui Language*, *Brief Annals of the Li Language*, *Brief Annals of the Qiang Language*, *Brief Annals of the Dongxiang Language*, and *Brief Annals of the Tu Language*. The writing of all the others is still in the process. Moreover, the *Mongolian-Chinese Dictionary*, *Tibetan-Chinese Dictionary*, *Concise Lisu-Chinese Dictionary*, *Uygur-Chinese Dictionary* and *Chinese-Dai Dictionary* have been published.

Theology Theology is a new branch of learning in China. Only in 1964 was a research institute on religion established. In February 1979, a national planning conference on religious study was called under the auspices of the World Religions Institute and a national plan for 1979-85 was then worked out. Principal areas of study and research include religious theories, Christianity, Islam, Buddhism, Taoism and Lamaism. A large number of books and articles have been written and published. Recent publications on religion include *Brief History of Christianity* (Vol. I), *Buddhism in the Sui and Tang Dynasties, Origin and Development of Buddhism in China, Origin and Development of Buddhism in India, Buddhism in China, The Three Major Religions in the World* (short guide), *History of Buddhism in China* (Vol. I), *Buddhism in the Song and Yuan Dynasties, Buddhist Scriptures of the Chan Sect* and *History of Taoism in China*. Subjects of study include not only the theory and the policy on religion but also the history of atheism in China and the rest of the world. Study is also conducted on the religions of minority nationalities. Reference books on the subject include *Dictionary of Religions, General Information on Religions in the World* and *21 Kinds of Buddhist Scriptures of Sakyamuni*, all of which will soon be published. Chinese translations of *Koran*, *Bible* and *Kitab al-Jami*

al-Sabih of al-Bukhari are also available. Other translations that will be in print include *History of Religion*
(in two volumes), *History of Islam, Buddhism in Japan*
and *Short History of Atheism in Europe.* Moreover, reference materials on the history of Christianity in China,
history of Buddhism in the world, history of Lamaism,
history of Taoism in China, history of Islamism in China
and history of Islam in the world have been collected and
compiled.

3. THE CHINESE ACADEMY OF SOCIAL SCIENCES AND ACADEMIC EXCHANGES WITH FOREIGN COUNTRIES

The Chinese Academy of Social Sciences The
Chinese Academy of Social Sciences is the highest academic institution and a comprehensive research centre in
social sciences in the People's Republic of China. Guided
by Marxism-Leninism and Mao Zedong Thought, its staff
conducts creative research on all aspects — social, political, economic and cultural — whether they be theoretical, historical, or current, so that China's sciences and
culture will continue to develop and flourish.

The duty of the academy is to study and determine
the directions and the tasks of the research institutes
under its jurisdiction, to organize them in their research
work, to strive to produce excellent research results, to
train highly competent scientists and researchers with
political consciousness, to undertake scientific research as
assigned by the Party and the state, to organize the research staff in various institutes to participate in the investigation and study of the important theoretical and

practical problems arising from socialist modernization, to promote academic exchanges and co-operation between its research institutes and the institutions of higher learning, departmental and local research organizations in various parts of the country, to take part in important international academic activities, and to encourage academic exchanges and co-operation between its social scientists and those in other countries.

In 1977, the Chinese Academy of Social Sciences came about as a result of reorganizing its predecessor — the Department of Philosophy and Social Sciences under the Chinese Academy of Sciences (established in 1955). The academy has now 32 research institutes and one research department, namely, Institute of Marxism-Leninism and Mao Zedong Thought, Institute of Philosophy, Institute of Economics, Institute of Industrial Economy, Institute of Agricultural Economy, Institute of Finance, Trade and Material Supplies, Institute of Technological Economy, Institute of Capital Construction Economy, Institute of Literature, Institute of Literature of Minority Nationalities, Institute of Foreign Literature, Institute of Linguistics, Institute of History, Institute of Modern History, Department of Contemporary History, Institute of Archaeology, Institute of World History, Institute of Law, Institute of Political Science, Institute of Nationalities, Institute of Sociology, Institute of Youth, Institute of Journalism, Institute of World Economy and Politics, Institute of World Religions, Institute of American Studies, Institute of the Soviet Union and Eastern European Studies, Institute of Japanese Studies, Institute of Western European Studies, Institute of West Asian and African Studies, Institute of South Asian Studies, Institute of Latin American Studies, and Institute of Information.

There are also Postgraduates Institute, Committee for Compiling Guo Moruo's Works, Committee for Compiling China's Seismic Data, *Social Sciences in China* magazine and Social Sciences Publishing House.

Hu Qiaomu is the Honorary President of the academy. Its President is Ma Hong and Vice-Presidents are Xia Nai, Qian Zhongshu, Liu Guoguang and Ru Xun.

Academic Exchanges with Other Countries Between 1978 and 1981, the Chinese Academy of Social Sciences signed 12 agreements on bilateral exchanges with academic institutions in Japan, the United States, Australia, Canada, Yugoslavia, Romania, the United Kingdom, Italy, Sweden, Holland and the Federal Republic of Germany. During the past four years, the academy has made contacts with its counterparts in more than 30 countries and regions.

As its academic exchanges with other countries grow, the academy has sent abroad and played host to an increasing number of scholars. In 1978, 21 Chinese scholars visited other countries in five groups while 49 foreign scholars came to China in 12 groups; in 1979, the number was 221 Chinese scholars in 52 groups, and 303 foreign scholars in 63 groups; in 1980, the number was 212 Chinese scholars in 78 groups, and 373 foreign scholars in 123 groups; and in 1981, the number was 282 Chinese scholars in 110 groups, and 557 foreign scholars in 157 groups. Altogether, the academy has in the past four years sent abroad 737 Chinese scholars in 245 groups, while playing host to 1,282 foreign scholars in 355 groups.

教育科学

《中国手册》编辑委员会编

＊

外文出版社出版

（中国北京百万庄路24号）

外文印刷厂印刷

中国国际书店发行

（北京399信箱）

1983年（32开）第一版

编号：（英）17050—167

00170

17—E—1664P